Information and Instructions

This shop manual contains several sections each covering a specific group of wheel type tractors. The Tab Index on the preceding page can be used to locate the section pertaining to each group of tractors. Each section contains the necessary specifications and the brief but terse procedural data needed by a mechanic when repairing a tractor on which he has had no previous actual experience.

Within each section, the material is arranged in a systematic order beginning with an index which is followed immediately by a Table of Condensed Service Specifications. These specifications include dimensions, fits, clearances and timing instructions. Next in order of arrangement is the procedures paragraphs.

In the procedures paragraphs, the order of presentation starts with the front axle system and steering and proceeding toward the rear axle. The last paragraphs are devoted to the power take-off and power lift systems. Interspersed where needed are additional tabular specifications pertaining to wear limits, torquing, etc.

HOW TO USE THE INDEX

Suppose you want to know the procedure for R&R (remove and reinstall) of the engine camshaft. Your first step is to look in the index under the main heading of ENGINE until you find the entry "Camshaft." Now read to the right where under the column covering the tractor you are repairing, you will find a number which indicates the beginning paragraph pertaining to the camshaft. To locate this wanted paragraph in the manual, turn the pages until the running index appearing on the top outside corner of each page contains the number you are seeking. In this paragraph you will find the information concerning the removal of the camshaft.

More information available at Clymer.com
Phone: 805-498-6703

Haynes Publishing Group
Sparkford Nr Yeovil
Somerset BA22 7JJ England

Haynes North America, Inc
859 Lawrence Drive
Newbury Park
California 91320 USA

ISBN 10: 0-87288-041-9
ISBN-13: 978-0-87288-041-2

SHOP MANUAL
ALLIS-CHALMERS

Models B-C-CA-G tractor serial stamped on top of transmission.

Models RC-WC-WF tractor serial stamped rear face of rear axle housing.

Models WD-WD45 tractor serial stamped rear face of transmission.

Model WD45 Diesel tractor serial number is located at left brake cover or on left rear side of transmission housing.

B-C-CA-RC engine serial stamped on top of left rear engine flange.

G engine serial stamped on side of left rear engine flange.

WC-WD-WD45-WD45 Diesel-WF engine stamped on left side of cylinder block.

BUILT IN THESE VERSIONS

Tractor Model	Tricycle Type Single Wheel	Tricycle Type Double Wheel	Axle Type Non-Adjustable	Axle Type Adjustable
B	No	No	Yes	Yes
C	Yes	Yes	No	Yes
CA	Yes	Yes	No	Yes
G	No	No	No	Yes
RC	Yes	Yes	No	No
WC	Yes	Yes	Yes	Yes
WD	Yes	Yes	No	Yes
WD45	Yes	Yes	No	Yes
WF	No	No	Yes	No
WD45D	Yes	Yes	No	Yes

INDEX (By Starting Paragraph)

CONDENSED SERVICE DATA
(WD 45 Diesel)

GENERAL

Tractor Model	WD 45 Diesel
Engine Make	Own
Engine Model	45D
Cylinders	6
Bore, Inches	3 7/16
Stroke, Inches	4 1/8
Displacement, Cubic Inches	230
Compression Ratio	15.5:1
Pistons Removed From?	Above
Main Bearings, Number of	7
Main Bearings, Adjustable?	No
Rod Bearings, Adjustable?	No
Cylinder Sleeves	Wet
Generator, Make	D-R
Starter, Make	D-R
Injection Nozzle, Make	Bosch
Injection Pump, Make	Bosch
Injection Pump, Model	PSB

TUNE-UP

Firing Order	1-5-3-6-2-4
Valve Tappet Gap, Inlet	0.010 H
Valve Tappet Gap, Exhaust	0.019 H
Inlet Valve Face Angle	45°
Exhaust Valve Face Angle	45°
Inlet Seat Angle	45°
Exhaust Seat Angle	45°
Injection Timing, Degrees BTC	21
Timing Mark Location	Flywheel

TUNE-UP *(Cont.)*

Tractor Model	WD 45 Diesel
Engine Low Idle RPM	600
Engine High Idle RPM	1975
Engine Loaded RPM	1625
Belt Pulley Loaded RPM	1460
PTO Loaded RPM	548

SIZES—CAPACITIES—CLEARANCES
(Clearances in Thousandths)

Crankshaft Journal Diameter	2.4975
Crankpin Diameter	1.998
Camshaft Journal Diameter, Numbers 1, 2 and 3	1.9985
Camshaft Journal Diameter, Number 4	1.2485
Piston Pin Diameter	0.9996
Valve Stem Diameter	0.3095
Compression Ring, Width	1/8
Oil Ring, Width	3/16
Main Bearings, Diameter Clearance	2.3-4.5
Rod Bearings, Diameter Clearance	1.5-3.5
Piston Skirt Clearance	4-5
Crankshaft End Play	2-7
Camshaft Bearing Clearance	2-4.6
Cooling System, Gallons	4 1/4
Crankcase Oil, Quarts (With Filter)	7

TIGHTENING TORQUE—Ft.-Lbs.

Connecting Rod Nuts	30-40
Cylinder Head Nuts	95-100
Flywheel Screws	95-105
Main Bearing Screws	125-135

MODEL WD45 DIESEL

CONDENSED SERVICE DATA
(All Non-Diesel Models)

TRACTOR MODEL	B (BE engine)	B-C CA	RC	WC WF	WD	G	WD45
GENERAL							
Engine Make	Own	Own	Own	Own	Own	Cont'l.	Own
Engine Model	BE	CE				N62	WD45
Cylinders	4	4	4	4	4	4	4
Bore—Inches	3¼	3⅜	3⅜	4	4	2⅜	4
Stroke—Inches	3½	3½	3½	4	4	3½	4.5
Displacement—Cubic Inches	116	125	125	201	201	62	226
Compression Ratio Non LP-Gas	4.92	6.2	5.75	5.5	5.75	6.5	6.45
Compression Ratio Non LP-Gas	4.67	5.75			5.0, 5.5	5.4	5.25
Compression Ratio Non LP-Gas		5.2, 4.7			4.5, 6.6		4.75
Compression Ratio LP-Gas							7.2
Pistons Removed From:	Above	Above	Above	Above	Above	Below (3)	Above
Main Bearings, Number of	3	3	3	3	3	2	3
Main Bearings, Adjustable?	(1)	(1)	(1)	(1)	(1)	No	(1)
Rod Bearings, Adjustable?	(1)	(1)	(1)	(1)	(1)	No	(4)
Cylinder Sleeves	Wet	Wet	Wet	Wet	Wet	None	Wet
Forward Speeds	3	*	4	4	4	3	4
Generator & Starter Make	D-R	D-R	D-R	D-R	D-R	D-R	D-R

*Band C – 3 speeds, CA – 4 speeds.

TUNE UP							
Firing Order	1, 2, 4, 3	1, 2, 4, 3	1, 2, 4, 3	1, 2, 4, 3	1, 2, 4, 3	1, 3, 4, 2	1, 2, 4, 3
Valve Tappet Gap	.010H	B&C .010H CA .012H	.010H	.012H	.012H	.012C	.012H
Valve Seat and Face Angle	45	45	45	45	45	45	45
Ignition Distributor Make	None	D-R	None	None	D-R	D-R	D-R
Ignition Distributor Model		1111735			1111745	1111708	1111745
Ignition Magneto Make	F-M	F-M	F-M	F-M	F-M	None	None
Breaker Gap	.020	.020	.020	.020	.020	.020	.020
Magneto Lag Angle	30°	30°	30°	30°	30°		
Magneto Impulse Trips	TC	TC	TC	TC	TC		
Retarded Timing Inches or Deg.	TC	TC	TC	TC	TC	3°B	TC
Full Advanced Timing Deg.	30°B	30°B	30°B	30°B	30°B	17°B	30°B
Mark Indicating:							
Retarded Timing		DC				Notch	DC
Full Advanced Timing	Fire	Fire	Fire	Fire	Fire		Fire
Mark Location—Flywheel, Fan Pulley	Fly.	Fly.	Fly.	Fly.	Fly.	F.P.	Fly.
Spark Plug—Make			———— Auto-Lite, AC or Champion ————				
Model for Gasoline			——— AN7 Auto-Lite, or 45 AC or J8 Champion ———				
Model for Low Octane			——— A11 Auto-Lite or 47 AC or J11 Champion ———				
Electrode Gap	.035	.032	.035	.035	.035	.025	.030
Carburetor Make Non LP-Gas			——— Marvel-Schebler and Zenith ———			Mar.-Scheb.	Mar.-Scheb.
Model (Marvel-Schebler)				TSX159	TSX159	TSV13	TSX464; 561
Model (Zenith)	61AJ7	161J7	161J7	161X7	161AX		
Float Setting (Marvel-Schebler)				9/32	9/32	¼	9/32
Float Setting (Zenith)	1 5/32	1 5/32	1 5/32	1 5/32	1 5/32		
Engine Low Idle rpm	475	475	475	475	475	475	475
Engine High Idle rpm	1850	*	1850	1575	1720	2100	1720
Engine Loaded rpm	1400	*	1500	1300	1400	1800	1400

SIZES—CAPACITIES—CLEARANCES							
(Clearances in thousandths)							
Crankshaft Journal Diameter—Front	2.2495	2.2495	2.2495	2.436	2.436	1.9995	2.436
Crankshaft Journal Dia., Center and Rear	2.2495	2.2495	2.2495	2.4775	2.4775	1.9995	2.4775
Crankpin Diameter	1.937	1.937	1.937	2.3745(2)	2.3745	1.4995	2.3745
Camshaft Journal Dia., Front and Center	1.7495	1.7495	1.7495	1.8745	1.8745	1.750	1.8745
Camshaft Journal Dia., Rear	1.7495	1.7495	1.7495	1.8745	1.8745	1.250	1.8745
Piston Pin Diameter	.8131	.8131	.8131	.9892	.9892	.5434	.9892
Valve Stem Diameter	11/32	11/32	11/32	3/8	3/8	5/16	3/8
Compression Ring Width	1/8	1/8	1/8	1/8	1/8	1/8	1/8
Oil Ring—Width	3/16	3/16	3/16	3/16	3/16	3/16	3/16
Main Bearings, Diam. Clearance	1-2	1-2	1-2	2-3	2-3	1.5-2.0	2-3
Rod Bearings, Diam. Clearance	1-2	1-2	1-2	1-3	1-3	1.5-2.0	1-3
Piston Skirt Clearance	2.5-4.5	2.5-4.5	2.5-4.5	2.5-4.5	2.5-4.5	2.0	2.5-4.5
Crankshaft End Play	1-5	1-5	1-5	3-7	3-7	3-7	3-7
Camshaft Bearing Clearance	2-4	2-4	2-4	2-3	2-4	3-4.5	2-4
Cooling System—Gallons	2	2	2	4	3½	1.6	3½
Crankcase Oil—Quarts	4	4	4	6	6	3½	6.0
Transmission—Quarts			4	4			
Differential—Quarts			6	6			
Transmission & Differential—Quarts	6	6 (CA) 8	10	10	17	8	17
Final Drive, Each—Quarts	.75	.75	.5	.75	1.75		1.75
Add for PTO and/or BP	1	1	1	1	1		1

(1) Bearings have shims which only control crush of shell insert. (2) On WC prior to serial 3665 crankpin diameter is 1.9995.
(3) Refer to text for alternative method of piston removal. (4) On engines prior W45FG1001, bearings have shims which only control crush of shell insert. On engines after W45FG1001, the bearings are shimless. *See paragraph 115.

FRONT SYSTEM—TRICYCLE AND AXLE TYPES

ADJUSTMENT OF PEDESTAL AND FRONT STEERING GEAR

Models C-CA-RC-WC-WD-WD45

1. C-CA SINGLE FRONT WHEEL. To adjust vertical spindle bearings, vary the shims (4—Fig. AC1), located between top of shaft and bearing cone. Radiator must be removed to obtain access to shims.

Backlash between bevel gears of front (auxiliary) steering unit should be .002, and is adjusted by varying the shims (9) after removing the radiator. Alternate the shims with gaskets to prevent oil leaks.

1A. C-CA ADJUSTABLE FRONT AXLE. The front support unit on these models houses the front steering gear and carries the pivot pin for the front axle. Internal construction, Fig. AC2, is same as single wheel models. To adjust, follow same procedure as outlined for single wheel C and CA in preceding paragraph.

1B. C-CA DUAL FRONT WHEELS. To adjust vertical spindle bearings, vary the shims (4—Fig. AC3) under cap screw (2) at top of shaft. To gain access to shims, remove the pedestal unit from the front support.

Backlash between bevel gears of front (auxiliary) steering gear should be .002. Adjust by varying the shims (9) being sure to alternate shims with gaskets to prevent oil leaks.

2. RC-WC-WD-WD45 SINGLE FRONT WHEEL. On RC-WC, and WD models prior serial WD25129, the wheel fork and vertical spindle are integral as shown in Fig. AC4. On later WD, after serial WD25128, and WD45, the wheel fork is flange bolted to the vertical spindle as shown in Fig. AC5.

Adjust vertical spindle bearings by varying shims (4) located at top of shaft underneath the bearing cone retainer.

To adjust steering wormshaft bearings, vary shims (46—Fig. AC6) to obtain free rotation with zero end play. Mesh of worm with sector is not adjustable.

Front wheel bearings are adjusted by varying the number of shims (40—Fig. AC4) located between the bearing retainer and wheel hub.

2A. RC-WC-WD-WD45 DUAL FRONT WHEELS. To adjust the vertical spindle shaft bearings to the desired .001-.003 end play, vary number

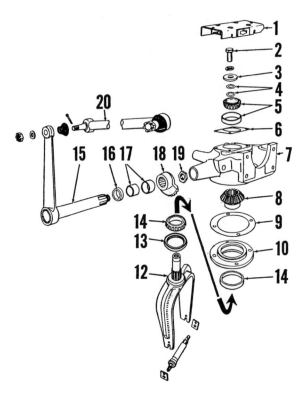

Fig. AC1—Models C and CA single wheel tricycle version front support and front steering gear assembly.

1. Radiator support
2. Cap screw
3. Washer
4. Shims
5. Bearing cone & cup
6. Gasket
7. Front support
8. Bevel pinion
9. Shims & gaskets
10. Bearing retainer
12. Vertical shaft & fork
13. Oil seal
14. Bearing cup & cone
15. Steering arm shaft
16. Oil seal
17. Bushings
18. Bevel segment gear
19. Retaining nut
20. Drag link

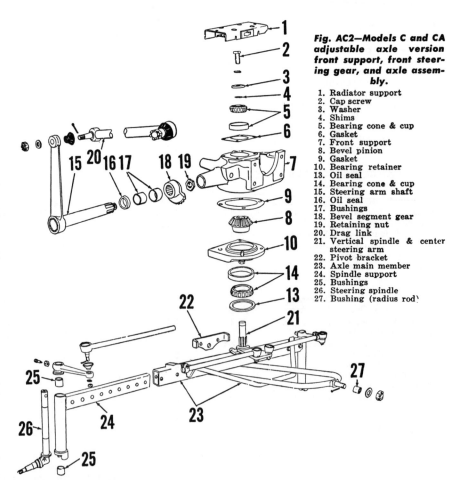

Fig. AC2—Models C and CA adjustable axle version front support, front steering gear, and axle assembly.

1. Radiator support
2. Cap screw
3. Washer
4. Shims
5. Bearing cone & cup
6. Gasket
7. Front support
8. Bevel pinion
9. Gasket
10. Bearing retainer
13. Oil seal
14. Bearing cone & cup
15. Steering arm shaft
16. Oil seal
17. Bushings
18. Bevel segment gear
19. Retaining nut
20. Drag link
21. Vertical spindle & center steering arm
22. Pivot bracket
23. Axle main member
24. Spindle support
25. Bushings
26. Steering spindle
27. Bushing (radius rod)

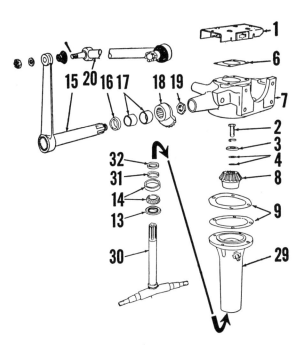

1.	Radiator support
2.	Cap screw
3.	Washer
4.	Shims
6.	Gasket
7.	Front support
8.	Bevel pinion
9.	Shims & gaskets
13.	Oil seal
14.	Bearing cone & cup
15.	Steering arm shaft
16.	Oil seal
17.	Bushings
18.	Bevel segment gear
19.	Retaining nut
20.	Drag link
29.	Pedestal
30.	Vertical spindle
31.	Seal retainer
32.	Seal

Fig. AC3—Models C and CA dual wheel tricycle version front support, pedestal, and front steering gear assembly.

of shims (60—Fig. AC8) located at bottom of spindle.

To adjust steering wormshaft bearings, vary the number of shims (46—Fig AC6) to obtain free rotation with zero end play. Mesh of worm with sector is not adjustable.

2B. WC-WD ADJUSTABLE FRONT AXLE. The vertical spindle bearings and front steering gear on these models, Figs. AC7 and AC9, are similar to the single wheel type, having the bolted-on type of fork. Adjustment procedure is also the same as outlined in paragraph 2.

OVERHAUL PEDESTAL & FRONT STEERING GEAR

Models C-CA-RC-WC-WD-WD45

3. C-CA SINGLE FRONT WHEEL. To disassemble pedestal and front gear unit, remove hood and radiator, and with tractor supported under torque tube, remove front wheel and horizontal spindle assembly from fork. Remove cap screw (2—Fig. AC1), washer (3), and shims (4) from top of vertical spindle shaft (12). Remove cap screws retaining bearing retainer (10) to front support (7), and bump shaft down through upper bearing cone and front support. Withdraw shaft, gear and bearing retainer assembly as a unit from bottom of front support. The need and the procedure for further disassembly will be determined by an inspection of the parts and by reference to Fig. AC1. Presized bushings (17) are supplied for steering arm shaft, and if carefully installed require no sizing after installation.

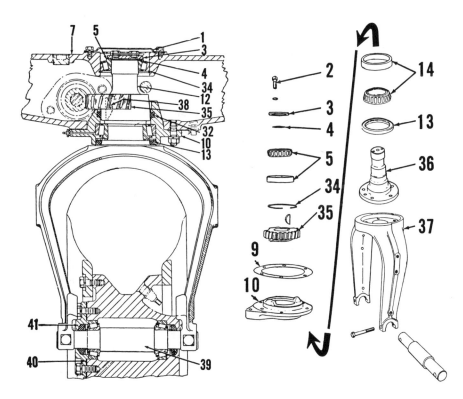

Fig. AC4—Models RC and WC single wheel tricycle version front support and main steering gear assembly. WD models prior to serial WD 25129 are similar.

1.	Cover	13.	Oil seal
3.	Washer	32.	Oil seal
4.	Shims	34.	Snap ring
5.	Bearing cone	35.	Sector
7.	Front support	38.	Sector key
10.	Bearing retainer	39.	Horizontal spindle
12.	Vertical shaft & fork	40.	Shims
		41.	Bearing retainer

Fig. AC5—Models WD & WD45 single wheel tricycle version main steering gear and wheel fork assembly. Effective on tractors after serial WD 25129. Refer to Fig. AC 4 for construction details prior to WD 25129.

2.	Cap screw	13.	Oil seal
3.	Washer	14.	Bearing cone & cup
4.	Shims		
5.	Bearing cone & cup	34.	Snap ring
9.	Gasket	35.	Sector
10.	Bearing retainer	36.	Vertical spindle
		37.	Wheel fork

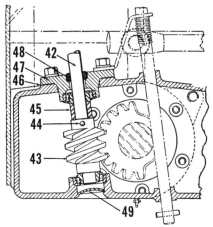

Fig. AC6—Models RC, WC, WD & WD45 single wheel, dual wheel, and adjustable axle type steering gear assembly.

42.	Worm shaft	46.	Shims
43.	Worm	47.	Bearing carrier
44.	Retaining pin	48.	Oil seal
45.	Spacer	49.	Expansion plug

3A. Assemble retainer, oil seal, lower bearing, bevel pinion gear and vertical shaft as a unit and install in front support after installing steering arm shaft and segment gear. Vary the number of shims (9) between retainer and front support to provide .002 backlash between bevel pinion and bevel segment gears. Alternate shims with gaskets to prevent oil leaks. Pinion gear and top of vertical shaft are punch marked and should be assembled as in Fig. AC10. Pinion gear and segment gear should be meshed so that steering arm is in position shown in Fig. AC11 when front wheel is in straight-ahead, trailing (castering) position and main steering gear is in mid-position. Adjust vertical shaft bearings by adding or removing shims (4—Fig. AC1) between top of shaft and bearing cone retaining washer.

Fork Mounted Wheel. To remove front wheel and horizontal spindle, remove cap screws and axle retainer. It will be noted that the cork seals are installed toward the outside, and the felt seal toward the inside.

3B. C-CA ADJUSTABLE FRONT AXLE. Except that vertical spindle in the front support terminates in a center steering arm instead of a wheel fork, the external parts are the same as on the single wheel models outlined in the preceding paragraph. Refer to Fig. AC2.

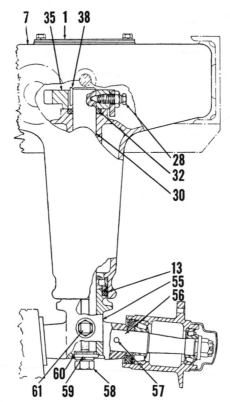

Fig. AC8—Models RC, WC, WD & WD45 dual wheel tricycle version front support, and main steering gear assembly.

1. Cover	55. Spindle block
7. Front support	56. Horizontal spindle
13. Oil seal	57. Spindle pin
28. Sector set screw	58. Cap screw
30. Vertical spindle	59. Washer
32. Oil seal	60. Shims
35. Sector	61. Set screw
38. Sector key	

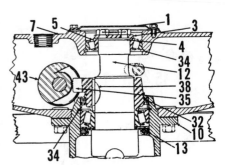

Fig. AC9—Models RC, and WC non-adjustable type axle version front support and main steering gear assembly.

1. Cover	12. Vertical shaft
3. Washer	13. Oil seal
4. Shims	32. Oil seal
5. Bearing cone & cup	34. Snap ring
7. Front support	35. Sector
10. Bearing retainer	38. Sector key
	43. Worm

3C. C-CA DUAL FRONT WHEELS. To disassemble pedestal and front steering gear, support tractor under torque tube and remove front wheels. Remove cap screws retaining pedestal (29—Fig. AC3) to front support (7) and remove pedestal. Remove cap screw (2), washer (3) and shims from top of vertical spindle shaft (30) and pull pinion gear (8) off shaft. Withdraw shaft, bearing cone and oil seal (13) from bottom of housing. The need and the procedure for further disassembly will be determined by an inspection of the parts and by reference to Fig. AC3. Presized bushings (17) are supplied for steering arm shaft, and if carefully installed require no sizing after installation. Removal of front support from tractor requires removing radiator core and detaching support from engine.

3D. Assemble steering arm shaft and segment gear before installing pedestal assembly. Pinion gear and vertical spindle shaft are punch marked and should be assembled as in Fig. AC10. Vary number of shims (4—Fig. AC3) between top of shaft and pinion gear retaining washer to remove all bearing play but permitting shaft to turn without binding. Install pedestal unit in front support varying the number of shims (9) to provide .002 backlash between pinion and segment gear. Pinion gear and segment gear should be meshed so that steering arm is in position shown in Fig. AC11 when front wheels are in straight-ahead, trailing (castering) position and main steering gear is in mid-position.

4. RC - WC - WD - WD45 WORM-SHAFT. To disassemble worm shaft, bump pin out of steering shaft universal joint and worm shaft and slide universal joint back. Remove steering shaft bearing bracket (47) and shims

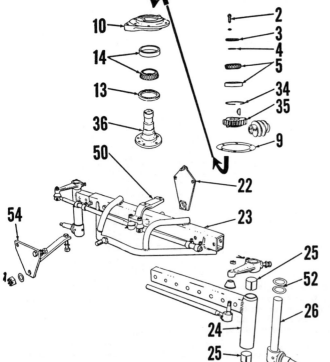

2. Cap screw	
3. Washer	
4. Shims	
5. Bearing cone and cup	
9. Gasket	
10. Bearing retainer	
13. Oil seal	
14. Bearing cone and cup	
22. Pivot bracket	
23. Axle main member	
24. Spindle support	
25. Bushings	
26. Steering spindle	
34. Snap ring	
35. Sector	
36. Vertical shaft	
50. Center steering arm	
52. Thrust washers	
54. Radius rod pivot bracket	

Fig. AC7—Models WD and WD45 adjustable axle version main steering gear, and axle assembly.

(46) from rear of front support and withdraw worm shaft and gear assembly. Shims (46) can be varied in number to adjust the wormshaft bearings.

4A. RC-WC-WD-WD45 VERTICAL SPINDLE. On single wheel models, the vertical spindle is supported in two roller bearings; on one such bearing in the dual wheel models. Refer to Figs. AC4, AC5, AC7, and AC8.

To R & R or renew the vertical spindle (30—Fig. AC8) and/or spindle taper roller bearing, it will be necessary to first support front end of tractor under frame rails and clutch housing. Remove pin from starting crank and withdraw crank from front support. On dual wheel type, remove cover (1)

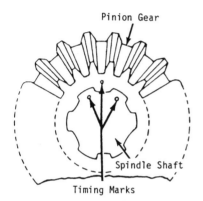

Fig. AC10—Correct position of pinion gear on vertical shaft is indicated by punch marks.

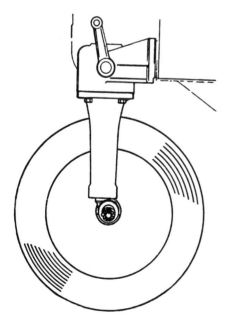

Fig. AC11—Models C and CA dual, and single wheel tricycle versions. Pinion gear and segment gear should be meshed so that steering arm is in position as shown when front wheel or wheels are in straight-ahead, trailing (negative castering) position, and main steering gear is in mid-position.

from front support and set screw (28) from sector (35). Using a suitable puller, remove sector and key. Withdraw vertical spindle and horizontal (axle) spindle unit from bottom of support. Vertical spindle taper roller bearing cone and/or cup, and oil seals (13 or 32), can be renewed at this time.

On single wheel models, Figs. AC4 & AC5, first remove retainer plate (3—Figs. AC4 & AC5) and shims (4) from top of vertical spindle. Remove stud nuts which hold retainer (10) to bottom of front support. Remove vertical spindle, bearings, sector and bearing retainer as an assembly from front support.

4B. When reassembling, adjust vertical spindle bearings of dual wheel models by varying number of shims (4—Fig. AC3). On single (fork mounted) wheel types, adjust bearings by varying the number of shims (4—Figs. AC4 & AC5).

4C. RC-WC-WD-WD45 WHEEL AXLE. On dual wheel type, the horizontal spindles (56—Fig. AC8) are individually available for service. Individual spindles can be removed from the spindle block after removing the front wheels, retaining pin (57) and adapting a combination puller.

Horizontal (axle) spindles and block (55) can be removed as an assembly after first removing the wheels, cap screw (58), washer (59), shims (60) and spindle block retaining set screw (61).

On single wheel type the spindle bearings are adjusted by varying the number of shims located between bearing retainer and wheel hub. Shims are shown at (40—Fig. AC4).

23. Radius rod (welded to axle member)
24. Spindle support
25. Bushings
26. Steering spindle
33. Steering arm
50. Center steering arm
51. Pivot, radius rod
53. Bushing, axle pivot
66. Dust shield
67. Tie rod & socket
68. Pivot pin

FRONT AXLE MEMBER

All Models

5. Adjustable, Figs. AC2, 7, and 13, or non-adjustable, Figs. AC14 & AC15 type front axles are either offered or are available on all models. On all such models, the axle main member complete with wheel spindles or knuckles and wheels can be removed from the tractor as a single unit. Exact procedure varies with the various models, but is self-evident after observing the actual tractor. On B, C, CA, G, WD and WD45, the radius rod is integral (by being welded thereto) with the axle main member.

SPINDLE (KNUCKLE) BUSHINGS

6. Steering spindle bushings and axle pivot bushings should be renewed if the diametral clearance exceeds .020. Steering spindle bushings usually require final sizing to provide .002-.005 clearance between spindle and bushing. To remove all wear in systems employing the Lemoine type spindles as in (26—Figs. AC2, 7, and 13), it may be necessary to also install new spindles. Recommended front wheel toe-in of 0-1/16 inch is adjusted by varying the length of tie rods.

RADIUS ROD

7. MODELS B-C-CA. Rear end of radius rod is pivoted in a removable bushing (27—Fig. AC2) in the front end of the torque tube (clutch housing). Bushing renewal required removal of front axle and integral radius rod.

7A. MODELS RC-WC. To remove radius rod, detach front end from axle and slide rear end off rear end of extended axle pivot pin.

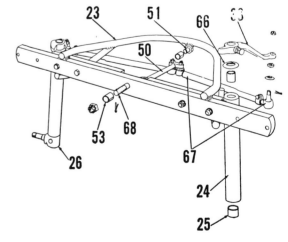

Fig. AC13—Model G adjustable axle has Lemoine type knuckles. Radius rod is welded to axle main member.

8. MODELS G-WD-WD45. Radius rod is integral with front axle and is supported in an unbushed bracket on WD and WD45 but the G is provided with a removable bushing.

9. MODEL WF. To remove radius rod, detach front end from axle and remove bolt attaching rear end to support plate on clutch housing.

PIVOT PIN (KING PIN)

10. MODEL B. Pivot pin is welded to front axle support plate and axle is retained to pivot pin by a cotter pin. On adjustable axle models, axle is equipped with a bushing for the pivot.

11. MODELS CA-WD-WD45. Pivot pin is welded to axle main member and rides in an unbushed bracket.

12. MODEL G. Pivot pin can be pressed out of axle main member. It is provided with a split type renewable bushing located in front frame.

13. MODELS RC-WC. Pivot pin can be removed, after removing radius rod, by disconnecting pivot pin bracket, removing set screw from axle, and driving pin rearward out of front support and axle.

14. MODEL WF. Pivot pin can be driven out of front support after removing cotter pin from behind axle and retaining pin from support.

MAIN STEERING GEAR

Service information in this section applies to models B, G, and WF, which have only a main steering gear, and to the main steering gear only on models C and CA. For information on the adjustment and overhaul of the front or auxiliary gear on models C, CA, RC, WD, and WD45 refer to paragraphs 1 through 4A.

B-C Gemmer Gear

(Refer to paragraphs 1, 1A, 1B, 3, 3A, 3B and 3C for model C front steering gear, and to paragraphs 22, 23, and 24 for Ross main gear.)

16. End play of worm shaft and sector shaft must be in adjustment before adjusting gear backlash. Disconnect drag link or remove ball arm (steering arm) to remove load from gear and permit locating mid-position.

17. WORM SHAFT END PLAY. First step in making this adjustment is to remove the fuel tank. To reduce play in worm shaft bearings, remove the steering wheel, loosen clamp (81—Fig. AC16) and slide tube (82) up column. Remove cap screws retaining

housing cap (80) to housing (on B and C work through tool box opening to remove two lower screws), and slide tube up and away from bearing retainer. Remove shims (78) until worm shaft has zero end play, but rotates freely. When re-positioning tube, be

sure upper end does not bind steering wheel.

18. SECTOR SHAFT END PLAY. It is advisable to remove the starter motor for better access to this adjustment. To adjust sector shaft (85) end play, place steering wheel about ⅛

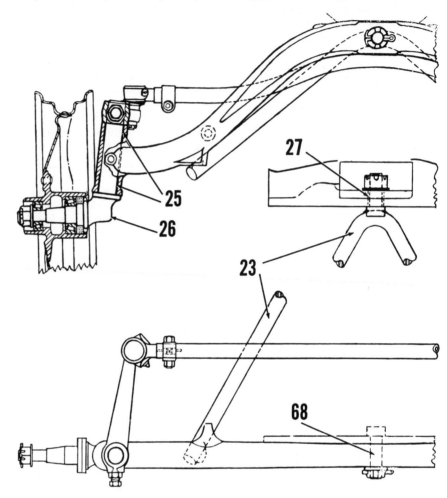

Fig. AC14—Model B non-adjustable axle version—front and top views.

| 23. Radius rod (welded to axle member) | 25. Bushings | 27. Bushing (radius rod) |
| 26. Steering spindle | | 68. Pivot pin |

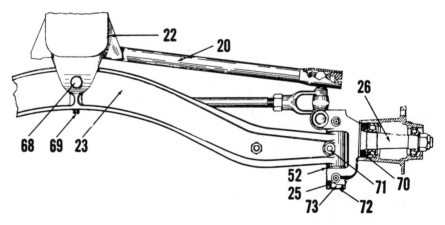

Fig. AC15—Models RC and WC non-adjustable axle version—front view.

20. Drag link	26. Spindle	70. Oil seal, hub
22. Front support	52. Thrust washer	71. Spindle lock stud
23. Axle member	68. Pivot pin	72. Expansion plug
25. Bushing	69. Set screw	73. Spindle pin

turn from either extreme position, loosen locknut (87) and turn adjusting screw (86) clockwise to reduce end play. Be sure housing nuts (92) and (93) are tight when making this adjustment. All end play in sector shaft should be eliminated but gear should not bind.

19. GEAR MESH. To adjust gear mesh, loosen column bracket clamp and locate mid-position of gear by turning steering wheel from one extreme position to the other then back half way. With gear in mid-position, loosen housing retaining nuts ¼ turn and eccentric sleeve jam nut (93) ½ turn. Slowly turn eccentric sleeve (95) clockwise while checking amount of gear backlash by feeling play in ball arm. (Turning eccentric sleeve clockwise moves steering gear housing down in relation to sector shaft housing bringing worm and sector into deeper mesh, thereby reducing gear backlash.) Adjust eccentric sleeve until play in ball arm is barly perceptible, when steering wheel is in mid-position. Make last adjustment of the eccentric sleeve in the clockwise direction. After mesh adjustment is completed, tighten eccentric sleeve jam nut, then housing nuts.

20. CENTRALIZATION. Check gear centralization by determining amount of play in ball arm when wheel is turned ⅓ turn to the right and ⅓ turn to the left of mid-position. If play is not the same on both sides of mid-position, loosen housing nuts and eccentric sleeve jam nut ¼ turn and centralize gear by turning eccentric rivet (88). If there is more play to the right, turn rivet clockwise and if more to the left, turn rivet counterclockwise. After gear has been centralized, readjust gear mesh with eccentric sleeve as previously described. Ball arm should be positioned on sector shaft so that steering gear is in mid-position when wheels are straight ahead.

21. OVERHAUL. Gear unit should be removed from tractor. Procedure is readily evident from an examination of Fig. AC16. When reassembling, install worm gear and shaft assembly and adjust worm gear bearings before assembling balance of steering gear. For complete adjustment procedure, refer to paragraphs 17 through 20.

B-C-CA-G-WF (Ross Gear)

(*Refer to paragraphs 1, 1A, 1B, 3, 3A, 3B and 3C for models C-CA front gear and to 16 through 21 for Gemmer main gear.*)

Before adjusting the gear, disconnect drag link or remove steering arm to remove load from gear and permit locating in mid-position.

22. WORM SHAFT (CAM) END PLAY. On B, C, and CA, first step in making this adjustment is to remove the fuel tank. To reduce play in cam

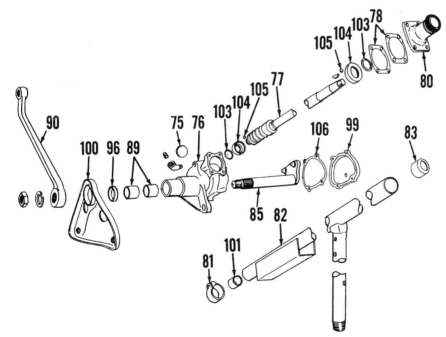

Fig. AC16—Gemmer steering gear assembly as used in early production B and C models.

75. Expansion plug	83. Bushing	91. Sector shaft housing
76. Housing	85. Sector & shaft	92. Housing nut
77. Worm & shaft	86. Adjusting screw	93. Jam nut
78. Shims & gaskets	87. Locknut	94. Conical lock ring
80. Housing cover	88. Eccentric rivet	95. Eccentric sleeve
81. Tube clamp	89. Bushing	96. Oil seal
82. Jacket tube	90. Steering arm (ball arm)	

Fig. AC17—Ross steering gear assembly (exploded view) as used in later and current production B, C, and all CA tractor models.

75. Expansion plug	83. Bushing	100. Housing bracket
76. Housing	85. Sector & shaft	101. Oil seal
77. Worm & shaft	89. Bushings	103. Snap ring
78. Shims & gaskets	90. Steering arm	104. Bearing cup
80. Housing cover	96. Oil seal	105. Bearing ball
81. Tube clamp	99. Side cover	106. Gasket
82. Jacket tube		

Fig. AC18—Ross steering gear installation on models B, C, and CA. Wormshaft bearings adjustment can be made after removing the fuel tank.

76. Housing	86. Adjusting screw
78. Shims & gaskets	87. Locknut
80. Housing cover	108. Tube clamp bolt
82. Jacket tube	

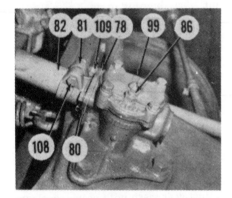

Fig. AC19—Ross steering gear installation on model G tractor.

78. Shims & gaskets	99. Side cover
80. Housing cover	108. Tube clamp bolt
81. Tube clamp	109. Housing cover
82. Jacket tube	cap screws
86. Adjusting screw	

(worm) bearings, remove steering wheel, loosen column jacket tube clamp Figs. AC17, AC18, AC19 and AC 20, and support clamp; then slide tube (82) up column. Remove cap screws retaining housing upper cover (80) to housing (on B, C, and CA, work through tool box opening to remove two lower cap screws). Remove shims (78) until shaft has zero end play, but rotates freely.

23. STUD (LEVER SHAFT) MESH. On B, C, and CA, it is advisable to remove the starting motor before attempting this adjustment. Locate mid-position of steering gear by turning steering wheel from one extreme position to the other, then back half-way. With gear in mid-position, loosen locknut (87—Fig. AC18 or AC20) and turn adjusting screw (86) clockwise until end play in lever shaft (cross shaft) is

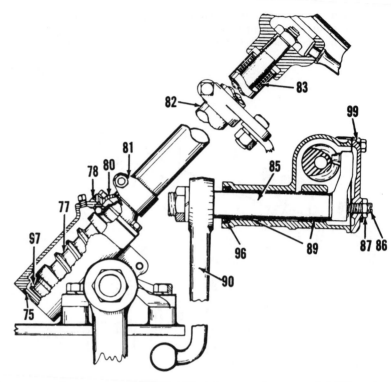

Fig. AC20—Ross steering gear assembly as used on later production model WF tractor.

75. Expansion plug	82. Jacket tube	89. Bushings
77. Worm & shaft	83. Bushing	90. Steering arm
78. Shims & gaskets	85. Sector & shaft	96. Oil seal
80. Housing cover	86. Adjusting screw	97. Snap ring
81. Tube clamp	87. Locknut	99. Side cover

barely perceptible. Be sure housing cover cap screws are all tight when making this adjustment. Shaft should have an increased amount of end play when gear is moved either way off the mid- or high point. Steering arm should be positioned on lever shaft so that steering gear is in mid-position when wheels are straight ahead.

On model G, steering wheel should be positioned on lever shaft splines in such manner that with front wheels pointing straight ahead, the cutaway section of steering wheel is located between the 12 and 4 o'clock positions.

24. OVERHAUL. On models B, C, and CA, removal of the gear unit requires removal of fuel tank, starting motor and its linkage, negative battery cable, steering wheel and shroud retaining bolts. After unbolting from torque tube, withdraw unit forward until it is free from shroud. The inner races for the worm shaft ball bearings are formed by the ground ends of the worm or cam itself. The lever shaft is supported on two renewable bushings, which are pressed into the gear housing. When reinstalling the jacket tube (82) the notch at its lower end should be up, to prevent grease leakage. Of the cap screws retaining the column shaft cover, the lower one or ones receive the grease sealing copper wash-

ers; the other cap screws are provided with lock washers.

RC-WC-WD-WD45

Steering gear on this model is located in front support. Refer to paragraph 4 and 4A for adjustment and repair data.

WF (Lavine Gear)

Before adjusting the gear, disconnect drag link or remove steering arm to remove load from gear and permit locating mid-position.

26. WORM SHAFT END PLAY. To reduce play in worm shaft bearings, remove shims (78—Fig. AC21) from between housing lower cover (80) and housing (76). All end play should be removed from worm gear but shaft should not bind.

27. GEAR MESH. Locate mid-position of steering gear, with drag link disconnected, by turning steering wheel from one extreme position to the other, then back half way. With gear in mid-position, loosen locknut and turn adjusting screw (86) clockwise to reduce end play in trunnion shaft (85). Be sure housing cover retaining cap screws are tight when making this adjustment. When end play in the trunnion shaft is reduced, drive nut (107) is meshed deeper into worm gear (77) and backlash is re-

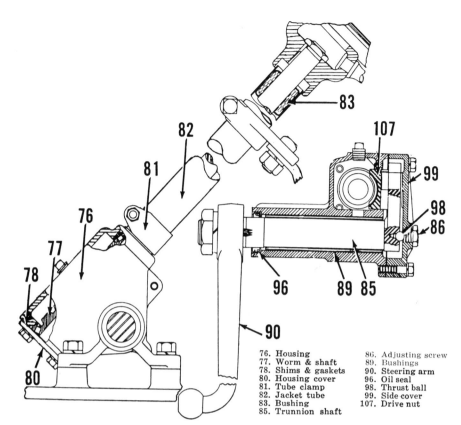

Fig. AC21—Lavine steering gear assembly as used on early production model WF tractor.

76. Housing	86. Adjusting screw
77. Worm & shaft	89. Bushings
78. Shims & gaskets	90. Steering arm
80. Housing cover	96. Oil seal
81. Tube clamp	98. Thrust ball
82. Jacket tube	99. Side cover
83. Bushing	107. Drive nut
85. Trunnion shaft	

duced. Steering arm should be positioned on trunnion shaft so that steering gear is in mid-position when wheels are straight ahead.

28. OVERHAUL. Gear unit should be removed from tractor. Procedure for disassembly is readily evident from an examination of Fig. AC21.

NON-DIESEL
ENGINE AND COMPONENTS

R & R ENGINE WITH CLUTCH

Models B-C-CA

60. Remove hood and disconnect fuel lines. Support tractor under torque tube. Disconnect steering drag link (also radius rod retaining nut if used), upper and lower radiator hoses, and light wires. With front axle and radiator and grille unit supported, remove four nuts which retain front support to engine. Move the front support radiator and pedestal or axle as a unit, away from the engine. Disconnect governor control, wires at coil, and generator, choke rod, and fuel line. With weight of engine supported by hoist, remove the four bolts retaining engine to torque tube and remove engine unit.

Model G

62. To remove engine from the tractor, drain cooling system and oil sump. Shut off fuel supply at tank connection. Remove hood, hood support (sheet metal cover from timing gear cover end of engine) and upper and lower radiator hose. Remove the drawbar guide. Disconnect starter control rod, choke rod, headlight wire, and battery cable at starter. Disconnect fuel line and remove throttle rod. The engine may be supported from the extreme ends of the fan shaft bearing supports, or by using a proper lift hook threaded into the number three spark plug hole (cylinders are numbered from timing gear end). Remove starting motor and bolts retaining engine to clutch housing. Engine may now be removed by pulling same forward to release clutch shaft from clutch plate.

Model RC

63. Remove hood and disconnect fuel lines, electrical connections, and controls. Remove radiator and crank pin located just behind front support and pull crank forward. Support engine in hoist and remove engine front support, flywheel cover and cap screws retaining engine to clutch housing.

Models WC-WD-WF-WD45

64. Remove hood and radiator. Disconnect fuel line, air cleaner hose and generator wire. Disconnect choke rod and throttle rod. Support engine with a hoist. Attach the lift bracket to two of the rocker shaft support studs. Remove universal joint retaining pin from steering worm shaft and slide steering shaft and universal joint rearward.

Remove two cap screws and two bolts retaining front engine support to frame rails and timing gear cover. Remove dust shield from front lower face of clutch housing. Remove cap screws retaining engine to clutch housing. Remove starting crank support and pull starting crank forward against spring pressure. Slide engine forward until free of clutch housing dowel pins. Raise front of engine until crankshaft pulley clears the axle front support and at the same time slide the engine forward until the clutch clears the clutch housing. Raise engine clear of frame rails and at the same time tip it toward left side of tractor so as to clear steering shaft.

Shims located between engine front support and lower side of timing gear cover should be varied to obtain proper engine alignment. The check for correct engine-to-clutch horizontal alignment should be made after the engine is bolted to the clutch housing and before the front support to timing gear cover cap screws are reinstalled.

R & R ENGINE AND CLUTCH
HOUSING AS A UNIT

Model G

65. Clutch housing and engine can be removed as a unit by following the engine removal procedure to the point of removing the engine to clutch housing retaining bolts. At this time disconnect clutch pedal rod from throwout bearing fork and unhook the brake return springs at the bottom of the radiator support. Use the same lift arrangement as for engine removal, except the balance point will be about 4 inches closer to flywheel end of engine. Remove clutch housing to transmission case retaining

bolts. Engine and clutch housing unit may now be removed by pulling same rearward to unmesh clutch shaft gear from mainshaft intermediate gear.

CYLINDER HEAD

All Except Model G

66. To remove cylinder head, remove hood and drain cooling system. Remove cylinder head cover (rocker lever cover) and rocker arms assembly. Unbolt carburetor from manifold and allow same to hang on air filter hose. Disconnect water outlet connection and oil connection to cylinder head. Remove cylinder head cap screws or nuts (also remove water manifold on WC, WD, WD45 and WF) and lift off head. When reinstalling, tighten head nuts or cap screws progressively and from the center outward. Retighten after engine has reached operating temperature. Tighten models B, C, CA, and RC cap screws to 60 foot pounds torque. Tighten models WC, WD, WD45 and WF ½ inch studs to 70 foot pounds and ⅜ inch studs to 25 foot pounds torque.

Model G

67. To remove cylinder head, first remove hood and support, and drain cooling system. Remove carburetor air intake, air cleaner, instrument panel, fan assembly, spark plug wire bracket, and upper radiator hose. Remove cylinder head retaining cap screws and nuts and lift off head.

When reinstalling, tighten head nuts progressively and from the center outward. Retighten after engine has reached operating temperature. Tighten the ⅜ inch cap screws to 53-57 ft. lbs. torque.

VALVES AND VALVE SEATS

All Except Model G

68. On B, C, and RC tappets should be set hot to .010 for inlet and ex-

haust valves. On CA, WC, WD, WD-45 and WF tappets should be set hot to .012 for both inlet and exhaust.

Inlet valves with a slightly larger head than exhaust valves seat directly in the cylinder head. However, inlet valve seats inserts are available for service. Oversize exhaust valve seat inserts are available for insert renewal. Both inlet and exhaust valves have a face and seat angle of 45 degrees. Desired seat width is 1/16 inch. Seats can be narrowed, using 30 degree and 60 degree cutters.

Engines in WD & WD45 after engine 299173 and engines 262650 through 263224, are factory equipped with Rotocap positive type valve rotators for the exhaust valves, Fig. AC23. These rotators are also available for service on the WC and WF engines. They require no maintenance but should be visually observed when engine is running to make sure that each exhaust valve rotates slightly. Renew the rotator element of any exhaust valve which fails to rotate.

Model G

69. Tappets should be set cold .012 inlet, and exhaust. Inlet and exhaust valves are not interchangeable. Inlet valves have round pin type spring locks while the exhaust valves, which are provided with Roto-valve release type rotators, have split cone type locks. All valves seat directly in the cylinder block and have face and seat angles of 45 degrees. Desired seat width is 3/64 inch—recut if wider than 1/16 inch. Seats may be narrowed, using 15 degree and 75 degree cutters.

A gap or end clearance of not less than .001 and not more than .006 between end of exhaust valve stem and bottom of Roto-valve cup should exist when valve is on its seat. Refer to Figs. AC24 and AC25.

VALVE GUIDES

All Except Model G

70. The cast iron presized shoulderless valve guides are pressed into the cylinder head with a piloted drift until scribe line mark on guide is flush with machined surface of valve spring seat. Refer to Fig. AC32, for WD & WD45 installation. Inlet guides are 7/32 inch

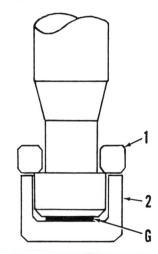

Fig. AC25—A gap or end clearance (G), of .001-.006 should exist in a Roto-valve installation when the valve is on its seat.

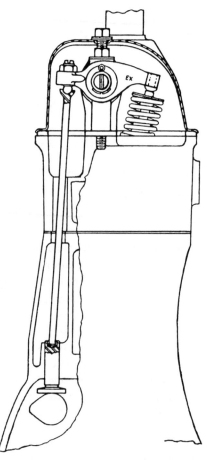

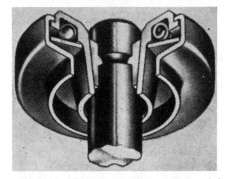

Fig. AC23—WD engines after engine serial 299173, and engines 262650 through 263-224 and WD45 are factory equipped with Rotocap positive type valve rotators for the exhaust valves. Rotators are also available for service on the WC and WF engines.

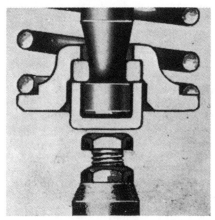

Fig. AC24—G tractor engine (N62 Continental) exhaust valves are provided with Roto-valve release type rotators.

Fig. AC26—Models B, C, CA, and RC engine valve system—end view.

longer than exhaust guides and extend farther into the ports. Desired clearance of the inlet and exhaust valve stems in the guides is .0025–.0045.

Model G

71. Inlet and exhaust guides are interchangeable. They are presized, but should be reamed after installation to .316 which will remove any localized high spots. Guides should be pressed or driven into the block, using a piloted drift .002 smaller than bore of guide, until top of each guide is 25/32 inch below gasket surface of the cylinder block. Desired clearance of inlet valve stems in guides is .002— of exhaust stems .0035. If inlet clearance exceeds .005, or exhaust clearance exceeds .007, renew the guides.

VALVE SPRINGS

72. Model G, inlet and exhaust valve springs are interchangeable. Each spring should require 13-17 lbs. pressure to compress it to a height of 1⅜ inches from its free length of 1 13/16 inches.

Spring can be removed without removing head by screwing adjusting screw down against lock and holding valve up with wire through spark plug hole.

On other models spring should require 33-39 lbs. pressure to com-

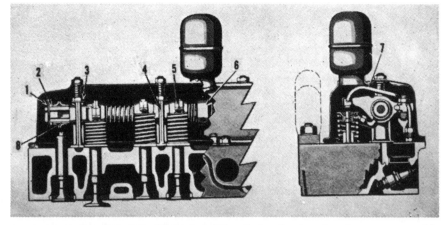

Fig. AC28—Models WC, WD, WD45 and WF cylinder head and rocker arm assembly showing exhaust valve seat inserts. Inlet valve seat inserts are available for service.

1. Retainer	4. Support	7. Oil tube
2. Washer	5. Rocker arm	8. Rocker bushing
3. Support stud nut	6. Oil collar	

press it to a height of 1¾ inches and 55-65 lbs. at a height of 1 5/16 inches. Springs which are rusted, distorted, or do not meet the foregoing pressure specifications should be renewed. On overhead valve models, springs may be renewed without removing the cylinder head after removing the rocker shaft.

CAM FOLLOWERS (Tappets)

All Except Model G

73. Mushroom type tappets (cam followers) operate directly in ma-

chined bores of the cylinder block. Tappets are furnished only in standard size and should have a clearance in block bores of .0005 to .002 with a maximum allowable clearance of .007 for service. Any tappet can be removed after removing the oil pan, rocker arms and shafts assembly, oil pump, timing gear cover and camshaft. Tappets should be adjusted to .010 hot for B, C, and RC; .012 hot for CA, WC, WD, WD45 and WF.

Model G

74. Barrel type tappets (cam followers) operate directly in the bores machined in the cylinder block. Tappets can be removed by first removing the valve spring which is accomplished by turning the adjusting screw down against the lock and holding the valve up with a wire through the spark plug hole. Now remove the tappet screw and lock element from the tappet. Tappets should be adjusted cold to .012 for both inlet and exhaust valves. Tappets are furnished only in standard size and should have .001 clearance in block bores.

VALVE ROCKER ARMS

All Except Model G

75. Rocker arms and shafts unit can be removed after removing the hood, valve cover, oil line from cylinder head to rocker arms and four shaft support retaining nuts and washers.

To disassemble the rocker arms and shafts assembly, remove hair pin type retainer or cotter pin and washer from ends of shaft. The valve contacting surface of rocker arms can be refaced, but the surface must be kept parallel to rocker arm shaft and original radius maintained. Desired clearance be-tween a rocker arm bushing and a new

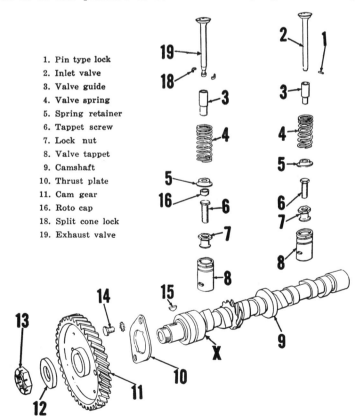

1. Pin type lock
2. Inlet valve
3. Valve guide
4. Valve spring
5. Spring retainer
6. Tappet screw
7. Lock nut
8. Valve tappet
9. Camshaft
10. Thrust plate
11. Cam gear
16. Roto cap
18. Split cone lock
19. Exhaust valve

Fig. AC27—Model G camshaft, valves and tappets. Note governor vent (X) in camshaft and items (16) and (18) which are used only on the exhaust valves.

shaft is .002-.003. If clearance exceeds .008, renew the rocker arm and bushing unit and/or shaft. Early production engines were equipped with forged rocker arms which have been discontinued and replaced by built-up, stamped steel type on all models. Rocker arm bushings are not available separately except for the discontinued forged type rocker arms.

The inlet valve rocker arm can be identified by a milled notch located on its upper surface between shaft and valve stem end. Reinstall rocker arm shafts with the oiling holes toward the cylinder head.

On B, C, CA, and RC, the grooved support stud shown in Fig. AC29 should be installed in rear stud hole. This stud acts as an oil passage to lubricate rocker shaft.

On all models be sure to install copper or composition washers between cover and cover nuts to insulate noise.

VALVE TIMING

All Except Model G

76. Valves are correctly timed when marked tooth of crankshaft gear is meshed with the correct mark on camshaft gear. If camshaft gear has two identical marks, the one opposite the keyway should be meshed with marked tooth of crankshaft gear, and the opposite one with the mark on the magneto gear. All "CE" engines and "BE" engines numbered BE12550 and higher have camshaft gear marked with a straight line and an "O" or a dot. On these engines, mesh the straight line of the camshaft gear with the similarly marked tooth on the

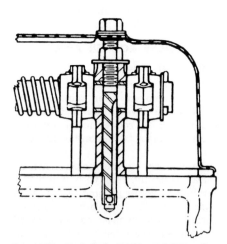

Fig. AC29—Models B, C, CA, and RC rocker arms. Rocker shaft rear support stud is grooved to provide oil passage to rocker arms.

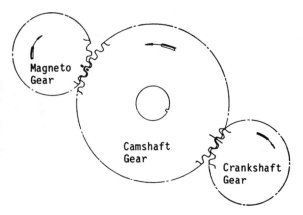

Fig. AC30 — Timing gear marks on all "CE" engines and "BE" engines numbered BE12550 and up. On "BE" engines prior to BE-12550, the camshaft gear has two identical marks; the one opposite the keyway should be meshed with the crankshaft gear.

crankshaft gear and the "O" mark with the "O" mark on the magneto gear. Refer to Fig. AC30.

Number one cylinder inlet valve opens at top center and exhaust valve closes 10 degrees after top center.

To check valve timing when engine is assembled, adjust number 1 inlet valve clearance to .014. Insert a .004 feeler gage between rocker arm and valve stem and rotate crankshaft until a slight drag is felt while trying to withdraw the feeler gage. At this time, the top center mark on flywheel should be in register with, or not more than ¼ inch either way from center of inspection port. Reset tappet gap to proper clearance. On flywheels which do not have a top center mark, the top center point can be established by measuring back from the "Fire" or "F" mark or by using the starter ring gear teeth as a basis for calculation. On the WD, which has a ring gear of 100 teeth, TC would be 8 1/3 teeth (3⅛″ on rim) after the "Fire" mark, as shown in Fig. AC46.

Model G

77. To check timing when engine is assembled, adjust both tappets of No. 4 cylinder to .020, then crank engine until No. 4 piston is at top center on the exhaust stroke. This point is indicated when the notch on crankshaft pulley is ⅛ inch past the center of the inspection port, located in the hood support as shown in Fig. AC53G. If valves are correctly timed, both valves will now be closed and it should be possible to rotate both tappets with the fingers.

TIMING GEAR COVER

Models B-C-CA

78. To remove cover, remove tractor hood and support tractor under torque tube. Remove radiator and axle or pedestal and front support as a single unit. Disconnect governor controls, and remove governor control housing. Remove crankshaft pulley, fan blades,

and fan belt. On some engines, the front-center oil pan retaining stud is welded to the angle iron on timing gear cover, necessitating removal of the oil pan assembly. Where the stud is not welded, cover can be renewed without removing the oil pan, but as a protection against subsequent oil leaks, it is advisable to always remove the oil pan. If oil pan is not removed, the oil pan gasket must be very carefully separated from the timing gear cover. Screws of three different lengths are used to retain the cover; be sure they are installed in the correct locations.

Model G

79. Although it is possible to renew crankshaft front oil seal without removing the timing gear cover, our suggestion is to remove the cover, which entails little additional work. To remove the timing gear cover, remove the hood, hood support and crank jaw. Disconnect carburetor link rod and governor spring. Remove pulley, using a puller and two ⅜ inch cap screws to attach same. Do not thread screws too far into pulley, as they may damage the cast timing gear cover. Remove the cover to cylinder block cap screws and withdraw the cover.

The seal which consists of an outer felt washer and an inner flanged ring of treated leather, should be installed with the lip of the inner element facing the timing gears.

Models RC-WC-WD-WD45-WF

80. It is possible to renew crankshaft front oil seal without removing the timing gear cover. The oil seal can be renewed after removing the hood, radiator, crankshaft pulley and front seal retainer. Loosen timing gear cover retaining cap screws to facilitate centering oil seal. Latest seals are of felt faced with neoprene.

To remove timing gear cover, first remove hood and radiator. Remove starting crank extension pin and loosen extension bracket. Loosen set screw retaining pulley to crankshaft and remove pulley. Disconnect throttle and governor linkage. Block up front end of engine and remove engine support. On RC, the front-center oil pan stud may be welded to the angle iron on timing gear cover; in which case it will be necessary to remove the oil pan. On WC, WD, WD45 & WF, it is not necessary to remove oil pan, but removal is recommended as a means of protecting against subsequent oil leaks where pan joins the bottom face of gear cover. Remove timing gear cover retaining cap screws and timing gear cover.

Refer to GOVERNOR section for overhaul of linkage.

Adjust camshaft end play after timing gear cover is installed by loosening thrust screw locknut (located on front face of cover) and turning thrust screw **in** until it contacts end of camshaft; then back the screw **out** ½ to ¾ turn.

TIMING GEARS
Models B-C-CA-RC

82. Camshaft gear is a press fit and keyed on the camshaft and should be removed, with the camshaft as a single unit. For camshaft removal procedure, refer to paragraph 85. Press new gear on until rear surface of gear hub is flush with front surface of camshaft front journal. Crank gear installation can be facilitated by a mild application of heat on the gear. For replacement of governor and magneto or distributor gear, refer to Governor, paragraph 116.

Model G

83. The drive consists of two helical gears. To remove the camshaft gear, first remove the timing gear cover as outlined in a preceding paragraph, and

Fig. AC33—Model G. Method of removing camshaft gear using two ⅜ inch cap screws and a suitable puller. Shaft should be bucked-up when reinstalling gear to shaft.

also remove the governor weight assembly from end of gear. Remove cam gear from shaft with a puller and two ⅜ inch cap screws as shown in Fig. AC33. Crankshaft gear can be removed with a suitable puller. When installing gears, remove oil pan or distributor and buck-up the camshaft at one of the lobes near front end of shaft with a heavy bar. Mesh the single mark on the crank gear between the double mark on the cam gear.

Models WC-WD-WD45-WF

84. Timing drive consists of three helical cut gears which includes the governor gear (drive for both governor and magneto). On WC, WD, WF and WD45 prior Eng. Ser. No. 45-48556 gear is attached to shaft by cap screws, and can be removed after removing the timing gear cover. The cap screw holes in the camshaft gear are unevenly spaced, permitting the gear to be installed to the camshaft in only one position. On later WD45 tractors, gear is pressed on the shaft.

Crankshaft gear is a press fit on the crankshaft, but can be removed when timing gear cover is off, by using a suitable puller. Reinstallation of this

gear can be facilitated by heating same in water or oil.

When reinstalling gears, mesh the single marked crankshaft gear tooth with a similar double mark on the camshaft gear. Refer to GOVERNOR section for governor and magneto or distributor drive gear data.

CAMSHAFT AND BUSHINGS
Models B-C-CA-RC

85. To remove camshaft, remove timing gear cover and oil pan. Remove cylinder head cover, rocker arms assembly and push rods. Hold tappets up to clear cams and withdraw camshaft and gear as a unit.

On RC, the steering gear pedestal or front support unit interferes, and must be removed to enable camshaft to be withdrawn. When reinstalling camshaft, be certain that oil passage (8—Fig. AC34) is clean. Oil passage in hollow camshaft delivers oil to crankshaft main bearings via the camshaft bushings, to the connecting rods and cylinders by throw holes (9) and to the timing gears through flutes in thrust plunger (11). Be sure thrust plunger operates freely in camshaft and that oil pump drive pin (6) is secured in camshaft.

Clearance between new camshaft and new camshaft bushings should be .002-.004 inch. Renew camshaft bushings or camshaft or both, if clearance exceeds .007 inch.

Bushings are supplied in Standard size and undersize of 0.0025. To renew camshaft bushings after camshaft has been removed, it is necessary to remove the flywheel and oil pump which necessitates engine removal.

The bushings can be driven out of block and new ones driven in. Drive rear bushing in from the front until rear edge of bushing is ¼ inch from

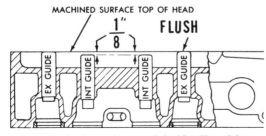
MACHINED SURFACE TOP OF HEAD
$\frac{1"}{8}$ FLUSH
EX GUIDE   INT GUIDE   INT GUIDE   EX GUIDE
NEW STYLE HEAD VALVE GUIDE INSTALLATION
Fig. AC32—Valve guide installation on WD engines after serial W 289000 and WD45. Inlet valve guides are installed so that the top end of the guide is ⅛ inch below the machined surface as shown. Exhaust valve guides are installed so that the top end is flush with the machined surface.

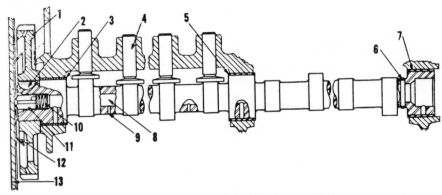

Fig. AC34—Models B, C, CA, and RC camshaft installation. Camshaft rotates on three renewable split type bushings. Camshaft end play is controlled by thrust plunger (11) and spring (10).

1. Cam gear
2. Camshaft gear key
3. Front bushing
4. Valve tappet
5. Center bushing
6. Oil pump drive pin
7. Rear bushing
8. Central oil passage
9. Oil hole for rod bearing lubrication
10. Thrust plunger spring
11. Thrust plunger
12. Oil slinger unit
13. Timing gear cover

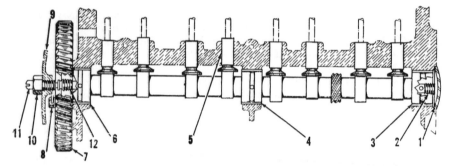

Fig. AC35—Models WC, WD, WD45 and WF camshaft installation. Camshaft rotates on three renewable split type bushings.

1. Expansion plug
2. Pipe plug
3. Rear bushing
4. Center bushing
5. Valve tappet
6. Front bushing
7. Cam gear
8. Gear retaining cap screw
9. Timing gear cover
10. Locknut
11. Thrust screw
12. Metered oil plug

rear of block. Drive center bushing in until it extends 1/16 inch beyond both ends of bearing boss. Drive front bushing in until front edge of bushing is flush with front face of block. Bushings are presized and should be driven in with a plug-type driver. The diameter of the plug should be 1.751 inches. Be sure oil hole in bushings registers with oil passage in cylinder block.

Journal Diameter (Mean) 1.7495
Running Clearance   .002-.004

Model G

86. To remove camshaft, first remove the camshaft gear, as outlined in preceding paragraphs. Remove the ignition distributor and cylinder head and either block up or remove, the valves and tappets. Remove camshaft thrust plate from front of cylinder block and withdraw the shaft.

Camshaft journals ride directly in 3 machined bores of the cylinder block. Shaft journal sizes are: Front and center 1.750; rear 1.250. Recommended clearance of journals in bores is .003 to .0045. The maximum permissible clearance is .007 and when it exceeds

this amount it will be necessary to renew the camshaft and/or the cylinder block, or both, or to make up and install bushings.

Camshaft end play is controlled by the thickness of the thrust plate (10—Fig. AC27) located behind the gear. Renew the thrust plate if the end play exceeds .007. Check the vent opening (X) in the camshaft directly behind the front camshaft journal. If the hole is plugged, it will be impossible to obtain satisfactory governor operation.

Models WC-WD-WD45-WF

87. To remove camshaft, remove timing gear cover, flywheel cover, oil pan and oil pump. Remove cylinder head cover, rocker arms assembly and push rods and remove cam gear from camshaft. Hold tappets up to clear cams and withdraw camshaft. When reinstalling camshaft, be sure oil passages are clean. Metering hole in plug (12—Fig. AC35) is discontinued in later WD and all WD45 models which have a solid plug. Adjust camshaft end play after timing gear cover has been installed, by loosening locknut,

turning thrust screw (11) in until it bottoms, then back ½-¾ turn.

Clearance between new camshaft and each of the three split type camshaft bushings should be .002-.004 inch. Renew camshaft bushings or camshaft or both if clearance exceeds .006 inch. Bushings are supplied in Standard size and undersize of 0.0025.

To renew camshaft bushings after camshaft has been removed, it is necessary to remove the flywheel which necessitates engine removal. Bushings can be driven out of block and new ones driven in. Drive rear bushing out towards the rear, driving expansion plug with it. Bushings are pre-sized and should be driven in with a plug-type driver. The diameter of the plug should be 1.875 inches. Be sure oil hole in bushing registers with oil passage in cylinder block. When installing expansion plug (1) in rear of block, be sure it seats tightly and seals completely.

Journal Diameter (Mean) 1.8745
Running Clearance   .002-004

ROD AND PISTON UNITS

Models B-C-CA-RC

88. Connecting rod and piston assemblies are removable from above after removing oil pan and cylinder head. Connecting rods are offset. Numbers 1 and 3 have the long part of the bearing toward the flywheel, and Numbers 2 and 4 have the long part toward the timing gears. Tighten connecting rod bolts to 35 foot pounds.

Model G

89. Piston and rod units are removed from below and only from the side opposite the camshaft. The lower portion of the cylinder bore is chamfered 45 degrees to permit installation of rings without using a ring compressor. Pistons and rods are installed with the rod cap correlation marks toward the camshaft. Replacement rods are not marked and should be installed with the oil spray hole at lower end of rod facing the camshaft, and with the piston slot facing opposite to camshaft side. Because of the close quarters in the crankcase end of the block, some mechanics claim time can be saved by reinstalling the assemblies from below, without any rings on the pistons. They then push the pistons up as far as they will go and assemble the rings to their piston grooves above the top face of the block. A ring compressor is then used to enter the rings into their cylinder bores. Tighten the connecting rod cap screws to 20-25 ft. lbs. torque.

Models WC-WD-WD45-WF

90. Piston and connecting rod assemblies are removed from above after removing cylinder head and oil pan.

Tighten the castellated connecting rod nuts to 70 ft.-lbs. torque and self-locking (lug lock) nuts to 40 ft.-lbs. torque.

ALUMINUM PISTONS. Pistons and rods are installed with the rod correlation marks facing the camshaft and the stamped arrow on top of each piston facing front of engine. The "T" slot of pistons should face opposite to camshaft side of engine. Replacement rods are not marked and should be installed with piston pin clamp screw facing the camshaft. Refer also to Fig. AC37.

IRON PISTONS. Piston pin bosses are of unequal length. Pistons 1 and 3 have the long pin boss toward the timing gears and Numbers 2 and 4 have the long pin boss toward the flywheel. Refer to Fig. AC38.

PISTONS, LINERS AND RINGS

Some of these engines are equipped with iron pistons and some with aluminum alloy pistons. Both types are cam ground. Iron pistons are equipped with a bushing in each pin boss; aluminum pistons are not bushed.

All Except Model G

91. Pistons are supplied in the standard size only and with variation in piston compression length for the three compression ratios.

The recommended piston skirt clearance is .0025-.0045, with a maximum clearance of .010 for service.

When assembling pistons to rods, and rods and piston units to crankshaft, do so as outlined in paragraphs 88 and 90.

With the piston and connecting rod assembly removed from the cylinder block, use a suitable puller to remove the wet type liner (sleeve). Before installing a new liner, clean all mating and sealing surfaces. The top of the liner should extend .002-.004 above the top of the cylinder block. If this standout is in excess of .004, check for foreign material under liner flange. Excessive standout will cause water leakage at cylinder head gasket. To facilitate installation of liners, use a lubricant (palm oil or vaseline) on the two neoprene sealing rings.

Some pistons are three ring type, and some are fitted with four rings. Recommended end gap of all rings is .007-.017. Recommended side clearance is .001-.0025.

On four ring pistons, the scraper type compression ring is installed in the third groove; on three ring pistons, in the second groove, always with the scraper groove down.

Model G

92. Aluminum alloy pistons are cam ground, and are provided with a .002 to .007 relief adjacent to both ends of the piston pin. Repair pistons are available in standard and in oversizes of .0035, .020, and .040. The recommended .002 skirt to cylinder clearance, is measured with a ½ inch feeler blade of .0015 thickness, which should require 10 to 15 lbs. pull to withdraw. Feeler should be inserted between skirt and cylinder block at right angles to the piston pin, with piston in upside down position as shown in Fig. AC39. Piston sizes are stamped on top of each piston.

Cylinders are unsleeved and according to the tractor manufacturer, are choke bored slightly (.0005) with the smaller diameter being near the top.

Standard size rings are used for service until cylinders are rebored. Rings are also furnished in oversizes of .020 and .040. However, the standard size rings used in factory production are not the same as the standard size rings provided for service, as will be seen from the table below:

	Production Standard Size	Service Standard Size
Compression Rings	2 tapered face, plain	Upper—chromium plated. Lower—tapered face with an expander
Oil Rings	One ventilated type	One ventilated, with expander

Recommended end gap of all rings is .007-.012. Recommended side clearance is .0015-.003 for compression rings; .001-.0025 for oil rings.

PISTON PINS

All Except Model G

93. The piston pins are furnished only in the standard size. Pin is locked in connecting rod.

Bosses of iron pistons are fitted with renewable bushings; aluminum pistons are not bushed. Connecting rods in B, C, CA, and RC are offset with respect to big end bearing; rods in WC, WD, WD45 and WF are straight. Iron pistons for WC, WD, and WF have pin bosses of unequal length; on aluminum pistons, bosses are of equal length.

As a check for pin fit in piston (piston at room temperature), the fit should be such as to prevent connecting rod from moving when piston and rod assembly is held in a horizontal position, yet a slight sudden movement of the piston will cause rod to move.

On B, C, CA, and RC, assemble pins to rods so that pin will be centered between bosses.

ALUMINUM PISTONS—WC-WD-WD45-WF. Pistons should be assembled to the connecting rods as follows: On all 4 units the connecting rod piston pin clamp screw should face the camshaft, and the arrow stamped on top of piston should point toward front, and the piston slot should face away from the camshaft. Assemble

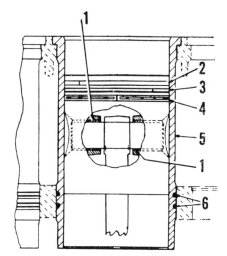

Fig. AC36—Piston and cylinder liner assembly as used in early production B and C engines, and all RC engines. Note iron pistons are equipped with piston pin bushings. CA engines are equipped with aluminum pistons which are not bushed for the piston pin. Piston pins are locked-in rod type for both cast iron and aluminum pistons.

1. Piston pin bushing
2. Piston ring
3. Scraper compression ring
4. Ventilated oil ring
5. Cylinder liner
6. Seal rings

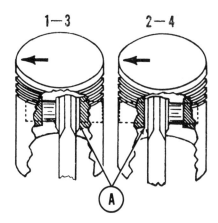

Fig. AC37—Models WC, WD, WD45 and WF piston pin to rod installation for models equipped with aluminum pistons. Piston pins are of the locked-in rod type. Numbers 1 & 3 units are assembled as shown at the left. Numbers 2 & 4 units are assembled as shown at the right.

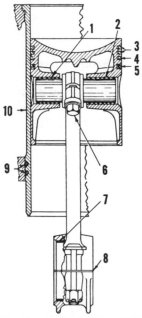

Fig. AC38—Models WC, WD and WF iron piston, liner and connecting rod assembly. Pistons 1 and 3 have long pin boss toward timing gears, and pistons 2 and 4 have the long pin boss toward the flywheel. Piston pins are of the locked-in rod type.

1. Pin bushing, short	6. Piston pin lock screw
2. Pin bushing, long	7. Rod bearing
3. Plain compression rings	8. Bearing shims (Not used on WD45 after W45FG1001)
4. Scraper compression ring	9. Seal rings
5. Ventilated oil ring	10. Cylinder liner

number one and three units so that the rod (A) is nearest the rear pin boss with end of pin slightly less than flush with same face of piston, as shown in Fig. AC37.

Assemble number two and four units in the opposite arrangement as shown.

IRON PISTONS—WC-WD-WF. Assemble pistons to connecting rods so that numbers 1 and 3 have the longer boss of the piston toward the timing gears and numbers 3 and 4 have the longer boss toward the flywheel. Refer to Fig. AC38.

Model G

94. The uncored (solid) .5434 diameter floating piston pins are retained in the piston bosses by snap rings and are available in standard and oversizes of .003 and .005. The bushing in the upper end of connecting rod has a very thin wall and if sized by reaming, do so only with a special fluted type and take very light cuts. Be sure oil hole in bushing registers with oil hole in top end of the connecting rod and be sure to clean same thoroughly after sizing the bushing. Pin should be fitted to a finger push fit (.0003 clearance) in the rod, and a thumb push fit in the piston.

Assemble pistons to rods to place split in aluminum piston on side opposite the camshaft.

CONNECTING RODS AND BEARINGS

All Except Model G (Shimmed Type)

95. Connecting rod bearings used in models B, C, CA, RC, WC, WD, WD45 prior engine serial W45FG1001 are of the slip-in type, provided with shims for a limited range of adjustment and are renewable from below.

When installing new bearing shells, be sure that the projection engages milled slot in rod and cap. Replacement rods are not marked and should be installed with the piston pin clamp screw facing the camshaft.

Bearings are available in .0025, .005, .040, .0425 and .45 undersize as well as standard for service

	B-C CA RC	WC WD WD45 WF
Crankpin diameter (mean)	1.937	2.3745
WC (prior to #3665)		1.9995
Running clearance (shimmed)	1-2	1-3
Side Clearance	4-7	5-8
Bolt Torque (ft.-lbs.)	35	*
Castellated with cotter pin	35	70
Self locking	35	40

*Tighten castellated connecting rod nuts to 70 ft-lbs. torque and secure with a cotter pin. On self-locking ("lug lock") nuts, torque nuts to 40 ft-lbs.

BEARING ADJUSTMENT. Connecting rod bearing shims which are provided for a limited range of adjustment do not fit between the bearing shell parting surfaces, but only between the bearing cap and its mating surface, as shown in Fig. AC40. These shims control the amount of crush or pinch placed on the shell and only indirectly control the bearing running clearance of 0.001-0.003.

Individual shim thickness of the factory supplied shim pack of four shims is 0.0025. Although Allis-Chalmers approves the removal of as many as two shims from each side as a means of reducing the running clearance, it is considered a better practice to install new undersize shells if new standard size shells do not restore the clearance to the recommended values.

In an emergency where new bearings are not available, the desired clearance can be obtained by reducing the height of the bearing shells. For each .002 reduction in height of a pair of shell inserts, remove a .0025 shim from each side. Removal of metal from the parting surface of shell (reducing) can be accomplished with fine emery paper and a flat surface, making an

occasional check on the bearing shell height to prevent removal of too much metal.

For new bearing shells and crankshaft installation, use the standard shim pack of four 0.0025 shims which will automatically provide the correct running clearance of 0.001-0.003 and the correct bearing crush of 0.0015.

Models WD-WD45 (Shimless Type)

95A. Connecting rod bearings used in WD45 tractors equipped with engines after serial W45FG1001 are of the shimless, non-adjustable, slip-in, precision type and are renewable from below after removing the oil pan and bearing caps.

Shimless type connecting rod bearings can be used in all WD and WD45 tractors if the shimless type connecting rods are used also.

When installing new bearing shells be sure that the projection engages the milled slot in both the rod and cap. Replacement rods are not marked and should be installed with the piston pin clamp screw facing the camshaft.

Bearings are available in .0025, .005, .040, .0425, and .045 undersize as well as standard.

Crankpin diameter (mean)	2.3745
Running clearance, shimless	1-3
Side clearance	5-8
Bolt torque (ft-lbs.)	*
Castellated with cotter pin	70
Self locking	40

*Tighten castellated connecting rod nuts to 70 ft-lbs. torque and secure with a cotter pin. On self-locking ("lug lock") nuts, torque nuts to 40 ft-lbs.

Model G

96. Connecting rod bearings are shimless, non-adjustable, slip-in, precision type, renewable after removing the oil pan and bearing caps. When installing new bearing shells, be sure that the projection engages milled slot in rod and cap. Bearings are available in .002 and .020 undersize, as well as standard.

Fig. AC39—Model G aluminum pistons are correctly fitted when 10-15 pounds pull is required to withdraw the .0015 feeler gage.

Crankpin diameter 1.499-1.500
Running clearance .... .0015-.002
  Renew if clearance
    exceeds .... .004
Side clearance .... .006-.010
  Renew if side clearance
    exceeds .... .014
Cap screw torque........20-25 ft. lbs.

CRANKSHAFT AND BEARINGS

All Except Model G

97. Shaft is supported on three slip-in type bearings provided with shims for a very limited range of adjustment. Bearings are renewable from below without removing crankshaft. Remove the engine to remove the crankshaft.

To remove the front bearing shell upper half on WC, WD, WD45, and WF only, it will be necessary to first remove timing gear cover, so as to lower shaft enough to disengage the shell from locating dowel (3—Fig. AC42). Loosen center and rear main bearing caps and remove front main bearing cap. Tap front main bearing upper half down and away from dowel pin and rotate shell around and off of crankshaft journal. All mains may be renewed on B, C, CA, and RC without disturbing the timing gear cover.

For new bearing shells and crankshaft installation, use the standard shim pack of four shims on each side which will automatically provide the correct running clearance and the correct bearing crush of .0015.

Bearings are available in .0025, .040, .0425 and .045 undersize as well as standard for service.

Oil wick (5—Fig. AC41 or AC42) for shaft pilot bearing lubrication which is inserted in the center of the rear main bearing journal can be renewed after removing engine clutch and oil pan. Insert wick from rear of journal and avoid stretching as a stretched wick will permit an excessive amount of oil to pass. The wick should be flush with front face of journal and extend

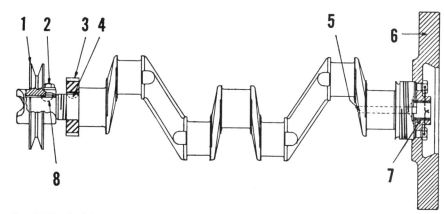

Fig. AC41—Models B, C, CA, and RC crankshaft assembly rotates on three slip-in type bearings which are provided with shims for a very limited range of adjustment.

| 1. Pulley | 3. Crankshaft gear | 5. Pilot bushing wick | 7. Pilot bushing |
| 2. Set screw | 4. Key | 6. Flywheel | 8. Pulley key |

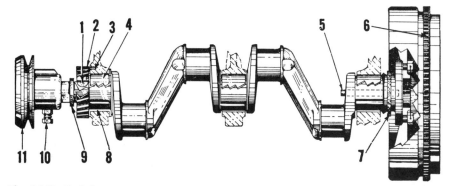

Fig. AC42—Models WC, WD, and WF crankshaft assembly. Front main bearing dowel (3) is used to prevent rotation of upper half of flanged front main bearing. Clutch pilot bearing wick (5) is not used in WD engines after W253200, and all WD45 engines.

| 1. Gear key | 3. Bearing dowel pin | 7. Rear oil seal | 9. Front oil seal |
| 2. Crankshaft gear | 5. Wick | 8. Lower bearing insert | 10. Pulley lock screw |

from the rear only enough to contact the clutch shaft. On WD beginning with engine W310920 and WD45 models the crankshaft has no oil wick or clutch shaft pilot bearing.

BEARING ADJUSTMENT. Crankshaft main bearing shims which are provided for a very limited range of adjustment do not fit between the bearing shell parting surfaces, but only between the bearing cap and its mating surface, as shown in Fig. AC40. Refer to paragraph 95 for further information.

Check the crankshaft journals for wear, scoring and out-of-round condition against the values listed below:

	B-C CA RC	WC WD WD45 WF
Journal Diameter:		
Front (mean)	2.2495	2.436
Center & rear (mean)	2.2495	2.4775
Running clearance	1-2	2-3
Crankpin		
diameter (mean) ...	1.937	2.3745
WC (prior #3665)...	——	1.9995
Renew or regrind if out-		
of-round more than..	.0025	.0025
Mains bolt torque		
(ft. lbs.)	80	85

Model G

98. Shaft is supported on two shimless, non-adjustable, slip-in, precision type main bearings, renewable from below, without removing the crankshaft. Bearing caps are installed with the numbers facing the camshaft. Crankshaft end play is controlled by shims (9—Fig. AC43), at number one journal and should be checked with all parts in place, including the crank jaw. Recommended end play is .003-

Fig. AC40—Models B, C, CA, RC, WC, WD, WD45 & WF main bearing shell installation. Note that shims do not fit between bearing shell parting surfaces. Connecting rod bearing construction and adjustment are similar.

X .0015 CRUSH

BEARING SHELL HEIGHT

SHIMS   SHIMS

.007. Adjust play when it exceeds .011. Both bronze thrust washers (5) should be installed with their beveled inside diameter facing the first crank throw. Install steel thrust plate (4) so that large surface contacts bronze washer. Shims (9) of .002 and .008 thickness are inserted between steel thrust washer and shoulder on crankshaft to obtain proper end play. To remove crankshaft it is necessary to remove engine, clutch, flywheel, rear oil seal, timing gear cover, and oil pan. Check the crankshaft journals for wear, scoring and out-of-round against values listed below:

Journal diameter 1.999–2.000
Crankpin diameter 1.499–1.500
Running clearance ...... .0015- .002
   Maximum .... .004
Renew or regrind if out-
   of-round .... .003
Main nut torque........75-85 ft. lbs.
Pal nut torque.......... 5-10 ft. lbs.
Undersize mains .... .002 & .020

CRANKSHAFT REAR OIL SEAL
Models B-C-CA-RC

99. To renew oil seal, separate engine from torque tube and remove flywheel and carefully lower the oil pan slightly, or remove it. The one piece seal is retained to rear surface of the cylinder block by a retainer and cap screws. Be sure oil seal contacting surface on crankshaft is smooth and true and oil seal is lubricated before assembly. Latest seals are of felt faced with neoprene. Copper washers should be installed under cap screw heads. If miner's wick is found at this point, replace it with copper washers.

Model G

100. Shaft rear oil seal is contained in the one piece main bearing closure plate which is bolted to the rear of cylinder block, as shown in Fig. AC45.

To renew the seal, first remove the engine and flywheel. Install seal in re-

tainer so that lip is towards timing gear.

Fig. AC45—Model G showing flywheel end of crankshaft. (1) Oil pump, (2) Oil pressure relief valve, and (3) Retainer for treated leather oil seal.

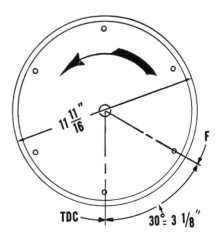

Fig. AC46—WD flywheel marks. WD45 flywheels are similar.

Models WC-WD-WD45-WF

101. Lower half of oil seal which is located in rear main bearing cap can be renewed after bearing cap is removed. Renewal of upper half of seal requires removal of engine, flywheel and seal retainer. Latest seals are of felt faced with neoprene.

Do not trim ends of seal as the seal will compress when bearing cap is tightened. Make certain oil seal contacting surface on crankshaft is smooth and true, and seal is lubricated before assembling.

CRANKSHAFT FRONT OIL SEAL

102. It is possible to renew the crankshaft front oil seal without removing the timing gear cover. If care is exercised, such a procedure will give satisfactory results. At any rate, it is advisable to loosen all of the timing gear cover screws when removing the old seal and to run the engine with them loosened so as to facilitate cen-

Fig. AC43—Model G crankshaft showing shims (9) for adjusting end play.

1. Starting crank jaw	4. Thrust plate	7. Bearing shell (upper)	10. Crankshaft gear
2. Crankshaft pulley	5. Thrust washer	8. Bearing shell (lower)	11. Oil slinger
3. Oil seal	6. Crankshaft gear key	9. Shims	12. Felt seal

Fig. AC44—Model G crankshaft rotates on two shim-less, non-adjustable, slip-in, precision type main bearings. End play is controlled by shims back of crank gear.

tering of the new seal to the crankshaft.

On model G, the inner flange ring element of the inner leather seal should be installed with the lip nearest the timing gears. Oil the leather and the outer felt elements before installing them. On B, C, CA, and RC, front face of oil seal should extend about ⅛ inch beyond the front of the timing gear cover. Latest seals are of felt faced with neoprene.

FLYWHEEL

103. On G, RC, WC, WD, WD45, and WF, R&R of flywheel requires removal of engine as outlined in paragraphs 62, 63 & 64. On B, C, & CA, flywheel may be removed after detaching (splitting) the clutch housing from the engine.

Flywheels used on models WD and WD45 equipped with a Rockford clutch are not interchangeable with the same models equipped with an Auburn clutch.

The starter ring gear can be renewed after detaching (splitting) the clutch housing from the engine without removing the flywheel from the crankshaft except on model G, which requires removal of flywheel from shaft. Reason for this is that on the model G, the ring gear must be installed to front edge rather than rear edge of flywheel. On model G, the bevel end of ring gear teeth should face toward the timing gears; on the others, the beveled teeth face rearward. However, the latest ring gears for B, C, and CA have straight teeth and can be installed either way.

On WD & WD45 with starter motor installed, distance from starter pinion to ring gear should be not more than ¼ inch. If greater than ¼ use either the later wider ring gear or relocate the starting motor locating hole in engine clutch housing.

FLYWHEEL TIMING MARKS
Models WD-WD45

104. Model WD flywheel carries only the magneto running timing mark "F", which is 30 deg. or 3⅛ inches before top center and is stamped on the flywheel rear face. To determine and position number one cylinder on top center, refer to Fig. AC46. Model WD45 flywheels are similarly marked with a running timing mark of "F" or "Fire" in addition to a "DC" mark. Flywheel marks are viewed through an inspection port located in clutch housing bottom cover.

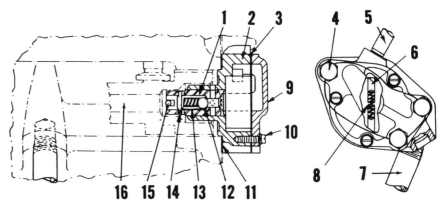

Fig. AC47—Models B, C, CA, and RC vane type oil pump which is mounted on rear face of engine block, and is driven by the camshaft. Install rotor blades so that the bevel edge faces toward direction of travel (anti-clockwise from drive end).

1. Pump shaft	7. Oil inlet pipe	13. Relief valve spring
2. Pump body	8. Rotor blade spring	14. Relief valve retaining pin
3. Cover gasket	9. Pump cover	
4. Retaining cap screw	10. Cover screw	15. Driving pin
5. Oil outlet pipe	11. Pump gasket	16. Central oil passage in camshaft
6. Rotor blade (2)	12. Relief valve ball	

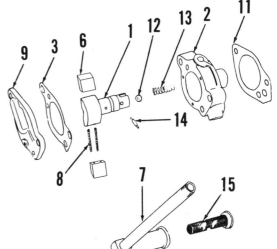

Fig. AC47A — Models B, C, CA, and RC vane type pump —exploded view. Install rotor blades so that the bevel edge faces toward direction of travel (anti-clockwise from drive end).

1. Pump shaft
2. Pump body
3. Cover gasket
6. Rotor blade (2)
7. Oil inlet pipe
8. Rotor blade spring
9. Pump cover
11. Pump gasket
12. Relief valve ball
13. Relief valve spring
14. Relief valve retaining pin
15. Inlet screen

OIL PAN (SUMP)

All Models

105. Method of removing the pan on all but WC-WD-WD45-WF is self evident, however some B-C-CA-RC engines have slightly longer cap screws installed at rear end of both sides of pan. Be sure these are installed in the correct location as they are used when the crankshaft rear oil seal retainer does not have steel inserts.

On WC-WD-WD45-WF it is necessary to remove engine front support and flywheel cover to provide assembling clearance at front end of pan. On these models make sure that corners of pan at each end of arch gasket are square and not depressed. If latter condition exists oil leakage may occur unless low spots are built up with solder or other suitable filler material.

OIL PUMP AND RELIEF VALVE
Models B-C-CA-RC

106. Oil pump can be removed after removing flywheel, by removing cap screws (4—Fig. AC47) retaining pump to rear surface of cylinder block. Free length of rotor blade springs should be 1 5/32 inches, and each should exert a pressure of 8-10 ounces when compressed to a length of ¾ inch. Rotor blades must be installed with bevel edge of blade toward direction of travel. Select gaskets (3) to obtain .001-.003 inch end play in rotor. If .004 inch end play is present with only one gasket installed, lap rear surface of pump body to reduce clearance.

The pressure relief valve is part of the rotor shaft unit and can be removed after driving out the retaining pin (14). Free length of valve spring should be 1⅛ inches and it should exert a pressure of 16-18 ounces when compressed to a length of ½ inch.

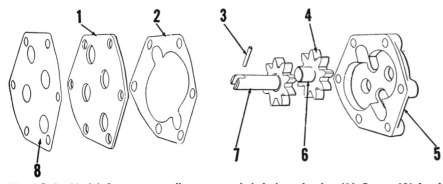

Fig. AC48—Model G gear type oil pump—exploded view, showing (1) Cover, (2) Lead gasket, (3) Drive pin, (4) Driven gear, (5) Pump body, (6) Driven gear shaft, (7) Drive gear.

Model G

107. Gear type pump is mounted at rear of the engine block and is driven by the camshaft. Pump removal requires removal of the engine, clutch and flywheel. The cap screws that attach oil pump to cylinder block also hold the assembly together. Check the pump for gear backlash, which should not exceed .005. Gear side clearance should not exceed .004. Desired clearance of the .4985 gear shaft in block and body bushing is .002-.003 and should not exceed .006. Renew the pump body if bushing is worn. The .501 diameter driven gear shaft is a press fit in pump body. A .007 lead gasket (2—Fig. AC48) is placed between pump body and cover and a paper gasket (8) is used between pump assembly and cylinder block.

The plunger type relief valve is located on the camshaft side of the engine rear mounting flange as shown in Fig. AC45. Relief valve spring should have a free length of 1 15/16 inches. Correct pressure is 15-20 lbs. controlled by inserting shims between spring and retaining plug.

Models WC-WD-WD45-WF

108. The gear pump, Fig. AC48A, which is driven by the camshaft, is mounted on the underside of cylinder block. Pump removal requires removal of oil pan.

Check the pump internal gears for backlash which should not exceed .015. The recommended diametrical clearance between gears and pump body should not exceed .008. Select gaskets (14) between oil pump body and cover to obtain .003-.005 clearance between pump gears and cover. Renew pump body and/or shaft or both if shaft to pump body clearance exceeds .008.

The piston type oil relief valve is located externally on right side of engine in the vicinity of the oil level dip stick. Oil pressure can be varied

by adding or removing shim washers between spring and retaining plug. Normal oil pressure is 12 lbs.

CARBURETOR

Non LP-Gas Models

110. Models B, C, CA, and RC are equipped with Zenith model 61AJ7 or model 161J7 carburetor. The G tractor comes equipped with a Marvel-Schebler TSV-13 carburetor. Models WC, WD, and WF are equipped with Zenith model 61AX7 or 161X7 or Marvel-Schebler model TSX159 carburetor. Model WD45 is equipped with a Marvel-Schebler TSX464 or TSX561 carburetor.

Float setting for Marvel-Schebler carburetor is 9/32 inch measured from nearest face of float to bowl cover gasket; for the Zenith carburetor, it is 1 5/32 inch measured from farthest face of float to bowl cover gasket. Turning the idle mixture screw towards its seat richens the mixture. Turning the power adjusting needle towards its seat leans the mixture. Idle speed stop screw should be adjusted to permit the engine to idle at 475-500 rpm.

Zenith carburetor outline numbers are not known and it will be necessary to refer to parts catalogs for calibration data. Marvel-Schebler calibration data are as follows:

TSV-13
Repair kit 286-834
Gasket set 16-656
Inlet needle and seat....233-570
Nozzle 47-293
Power jet 49-505

TSX-159
Repair kit 286-685
Gasket set 16-592
Inlet needle and seat....233-536
Idle jet or tube.......... 49-101L
Nozzle 47-303
Power jet 49-208

TSX-464
Repair kit 286-1037
Gasket set 16-654
Inlet needle and seat....233-543
Idle jet or tube.......... 49-101L
Nozzle 47-343
Power jet 49-364

TSX-561
Repair kit 286-1037
Gasket set 16-683
Inlet needle and seat....233-543
Idle jet or tube.......... 49-101L
Nozzle 47-343
Power jet 49-364

Fig. AC48A—Models WC, WD, WD45 and WF gear type oil pumps. Select gaskets (14) to obtain 0.003-0.005 clearance between pump gears and cover.

1. Helical drive gear
2. Retaining pin
3. Retaining screw
4. Pump body
5. Pump shaft
6. Thrust collar
   (Not used on WD45)
7. Idler gear shaft
8. Idler gear
9. Screen retainer
10. Oil screen
11. Cover screw
12. Pump cover
13. Pump gear key
14. Cover gasket
15. Pump gear
16. Thrust collar pin

LP-GAS SYSTEM

Fig. AC49—Ensign LP-Gas rgulator, economizer and combination LP-Gas and gasoline carburetor installation on a model WD45 tractor.

1. Water inlet	8. Vacuum connection	12. Gasoline fuel inlet
2. Water outlet	9. LP-Gas partial load	13. Gasoline inlet valve
3. LP-Gas idle connection	mixture adjustment	shut-off
4. LP-Gas idle mixture	10. LP-Gas economizer	14. Carburetor throttle to
adjustment	assembly	governor linkage
6. LP-Gas fuel inlet	11. Gasoline idle mixture	15. Gasoline load mixture
7. Balance line	adjustment	adjustment

CARBURETOR

Model WD45 LP-Gas

The model WD45 tractor is available with an LP-gas system designed and built by Ensign Carburetor Co. Like other LP-gas systems, this system is designed to operate with the fuel tank not more than 80% filled.

The LP-gas system consists of an Ensign model Kg1, which is a straight LP-gas carburetor Fig. AC49A, or a model Kgn1, Fig. AC49B, which is a combination gasoline and LP-gas carburetor. Both carburetor models are fitted with an Ensign model 3069 dry gas economizer, Fig. AC50. The vaporizer and pressure regulating device, Fig. AC51, is an Ensign model W unit. Fuel is filtered by a model 6259 Ensign filter, Fig. AC50A.

111. MODEL Kg1 CARBURETOR. The model Kg1 straight LP-gas carburetor and model W regulator combination have 3 points of mixture adjustment, plus an idle stop screw. Refer to Figs. AC49A & AC51.

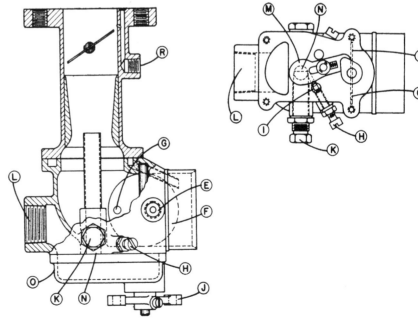

Fig. AC49A—Model Kg1 Ensign carburetor, which is designed to operate on LP-gas, is used on some model WD45 tractors. An Ensign model 3069 dry gas economizer (Fig. AC50) is used in lieu of the fuel adjusting load screw (K).

E. Balance line connection	J. Choke lever	M. Cam lever
F. Choke disc	K. Economizer location	N. Main orifice
H. Starting screw	L. Gas inlet	O. Cover plate

111A. IDLE STOP SCREW. Idle speed stop screw on the carburetor throttle should be adjusted to provide a slow idle speed of 475 crankshaft rpm or 429 belt pulley rpm.

111B. STARTING SCREW. Immediately after the engine is started, place the throttle in the wide open position and the choke in the fully closed position. Rotate starting screw (H—Fig. AC49A) until the highest engine speed is obtained. A slightly richer adjustment (counter-clockwise until speed drops slightly) may be desirable for a particular fuel or operating condition. Average adjustment is one turn open.

111C. IDLE MIXTURE SCREW. With the choke open, engine warm and idle stop screw set, adjust idle mixture screw (K—Fig. AC51) which is located on regulator until best idle is obtained. An average adjustment is ¾ of a turn open.

111D. LOAD SCREW OR ECONOMIZER. Refer to Fig. AC50. The effect of the load screw or economizer (located on carburetor) is to provide adjustable control of the partial or light load mixture by means of adjusting screw (A). The richer mixture required for maximum power is adjusted by varying the depth to which the entire economizer body is screwed into the fuel passage of the carburetor. The economizer body thus becomes the load screw.

In partial or light load operation, the economizer plunger (6) is moved into the main fuel passage by the high manifold vacuum applied behind diaphragm (3). Restricting or partially closing off the main fuel passage provides a leaner part throttle mixture. When manifold vacuum falls below 4-6 inches Hg., the economizer spring retracts the plunger and provides a richer power mixture.

111E. ADJUSTMENT WITHOUT LOAD. Carefully adjust the idle mixture as outlined in paragraph 111C. Remove the vacuum line from the economizer and plug the line or line connection at the inlet manifold. Disconnect rod from carburetor throttle lever and rotate idle stop screw to produce a high idle rpm of 1720. With economizer partial load adjusting screw (A) backed out against the stop, loosen economizer body locknut and rotate the entire unit until maximum rpm is obtained. Note this position; then rotate economizer body in an opposite (counter-clockwise) direction until rpm begins to fall. Rotate the entire unit to the mid-point of these positions and tighten the locknut. Unplug the manifold connection and reconnect the vacuum line.

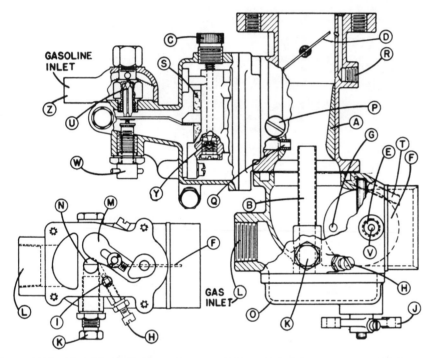

Fig. AC49B—Model Kgnl Ensign carburetor, which is designed to operate on LP-Gas or gasoline, is used on some model WD45 tractors. Fuel adjusting load screw (K) is replaced with an Ensign model 3069 dry gas economizer. Thumb screw (W) is used to close off the gasoline inlet valve.

A. Venturi	L. Gas inlet	R. LP-Gas idle fuel inlet
B. Nozzle	M. Cam lever	S. Gasoline float
C. Gasoline main load adjusting screw	N. Main orifice	T. Balance line pitot tube
	O. Cover plate	U. Gasoline inlet valve
F. Choke disc	P. Gasoline idle mixture adjusting screw	V. LP-Gas pitot tube
H. LP-Gas starting screw		W. Thumb screw
J. Choke lever	Q. Venturi nozzle	Y. Orifice, load screw
K. Economizer location		Z. Gasoline fuel inlet

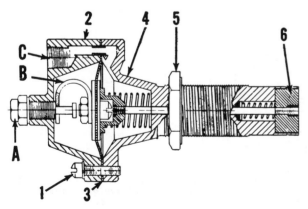

Fig. AC50—Ensign model 3069 dry gas economizer as used with the Ensign Kgl or Kgnl carburetor on the model WD45 engine.

A. Adjusting screw (partial load)
B. Adjusting screw stop
C. Vacuum connection
1. Cover attaching screws
2. Cover
3. Diaphragm
4. Body assembly
5. Locknut
6. Plunger

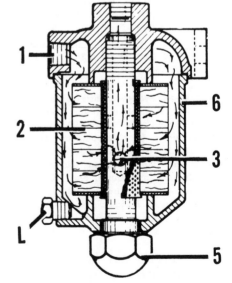

Fig. AC50A—LP-Gas filter used on model WD45 engine.

L. Drain plug
1. Fuel inlet
2. Filter cartridge
3. Outlet passage
5. Stud nut
6. Filter bowl

Rotate the partial load mixture adjusting screw (A) in a clockwise direction until the engine rpm begins to fall; then lock the adjustment.

111F. ADJUSTMENT WITH ANALYZER AND VACUUM GAGE. Remove the vacuum line from the economizer and plug the line or line connection at the inlet manifold. In this method, the engine is operated with the carburetor throttle wide open and with sufficient load on the engine to hold the rpm to maximum operating speed (1400 rpm) or 300 to 500 rpm slower than the maximum. With the economizer partial load adjusting screw (A) backed out against the stop, rotate the entire economizer body to give a reading of 12.8 on the analyzer with a gasoline scale or 14.3 on the analyzer with an LPG scale.

111G. Reconnect the vacuum line, check the part throttle (partial load) mixture adjustment by reducing both the opening of the throttle valve and the load until a manifold vacuum of 10-13 inches is obtained at the same rpm as used in making the full load mixture adjustment. Rotate the economizer adjusting screw (A) in a clock-

wise direction until the analyzer reads 13.8-14.5 on the gasoline scale or 14.9-15.5 on the LPG scale.

112. MODEL Kgn1 CARBURETOR. Refer to Fig. AC49B. The LP-gas portion of the Ensign model Kgn1 carburetor (gasoline and LP-gas) and model W regulator have 3 points of mixture adjustment, plus an idle stop screw. Follow the adjusting procedures outlined in paragraphs 111A through 111G when adjusting the LP-gas section of the combination carburetor.

112A. The gasoline section of this carburetor has 2 points of mixture adjustment, plus an idle stop screw. The idle stop screw is adjusted to provide an idle speed of 475 crankshaft rpm or 429 belt pulley rpm. The idle mixture screw (P) and load screw (C) are adjusted in a manner similar to that of any other gasoline carburetor. Turning the idle mixture screw towards its seat richens the mixture. Turning the load screw towards its seat leans the mixture.

112B. The carburetor float is adjusted so that the top of the float indexes with the first or top line which

is cast in the interior of the float chamber. The second or bottom line is used to adjust the float when the system uses a fuel pump.

LP-GAS FILTER

113. Filters used with the LP-gas systems are subjected to pressures as high as 150 psi and should be able to stand this pressure without leakage. Unit should be drained periodically at the blow off cock (L—Fig. AC50A). When major engine work is performed, it is advisable to remove the lower part of the filter, thoroughly clean the interior and renew the felt cartridge of same if not in good condition.

LP-GAS REGULATOR

114. HOW IT OPERATES. In the Ensign model W regulator, fuel from the supply tank enters the regulating unit (A—Fig. AC51) at a tank pressure of 25-80 psi and is reduced from tank pressure to approximately 4 psi at the inlet valve (C) after passing through the strainer (B). Flow through the inlet valve is controlled by the adjacent spring and diaphragm. When the liquid fuel enters the vaporizing chamber (D) via the valve (C) it ex-

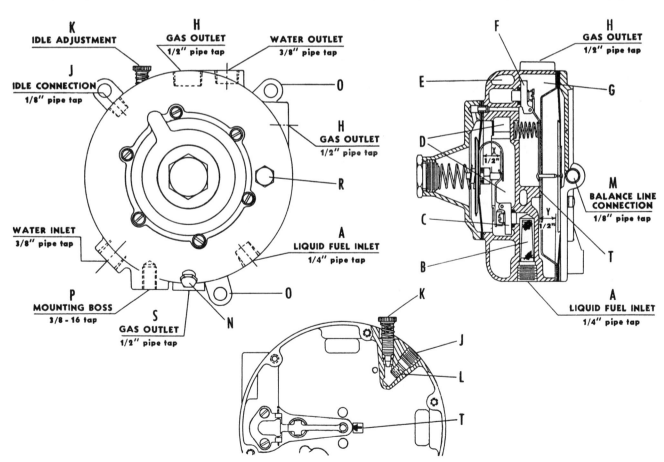

Fig. AC51—Model W Ensign LPG regulator used on the WD45 engine—assembled view.

A. Fuel inlet
B. Strainer
C. Inlet valve

D. Vaporizing chamber
E. Water jacket
F. Outlet valve

G. Low pressure chamber
H. Gas outlet
J. Idle connection

L. Orifice (idling)
M. Balance line connection
T. Boss or post

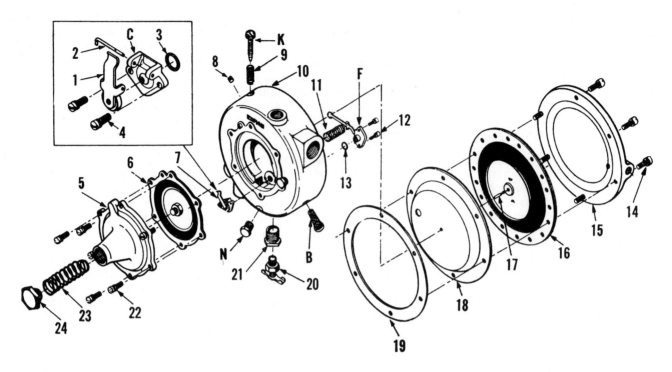

Fig. AC51A—Model W Ensign LPG regulator used on WD45 engine—exploded view.

B. Strainer	5. Regulator cover	11. Outlet diaphragm spring	18. Partition plate
C. Valve seat	6. Inlet pressure diaphragm	13. "O" ring	19. Partition plate gasket
F. Outlet valve assy.	7. Inlet valve assembly	15. Back cover plate	20. Drain cock
K. Idle adjusting screw	8. Bleed screw	16. Outlet pressure diaphragm	21. Reducing bushing
1. Inlet diaphragm lever	9. Idle screw spring	17. Push pin	23. Inlet diaphragm spring
2. Pivot pin	10. Regulator body		24. Spring retainer
3. "O" ring			

pands rapidly and is converted from a liquid to a gas by heat from the water jacket (E) which is connected to the coolant system of the engine. The vaporized gas then passes at a pressure slightly below atmospheric pressure via the outlet valve (F) into the low pressure chamber (G) where it is drawn off to the carburetor via outlet (H). The outlet valve is controlled by the larger diaphragm and small spring.

Fuel for the idling range of the engine is supplied from a separate outlet (J) which is connected by tubing to a separate idle fuel connection on the carburetor. Adjustment of the carburetor idle mixture is controlled by the idle fuel screw (K) and the calibrated orifice (L) in the regulator. The balance line (M) is connected to the air inlet horn of the carburetor so as to reduce the flow of fuel and thus prevent over-richening of the mixture which would otherwise result when the air cleaner or air inlet system becomes restricted.

114A. TROUBLE SHOOTING. The following data should be helpful in shooting trouble on the WD45 LP-Gas tractor.

114B. SYMPTOM. Engine will not idle with idle mixture adjustment screw in any position.

CAUSE AND CORRECTION. A leaking valve or gasket is the cause of the trouble. Look for a leaking outlet valve caused by deposits on valve or seat. To correct the trouble, wash the valve and seat in gasoline or other petroleum solvent.

If the foregoing remedy does not correct the trouble check for a leak at the inlet valve by connecting a low reading (0-20 psi) pressure gage at point (R—Fig. AC51). If the pressure increases after a warm engine is stopped, it proves a leak in the inlet valve. Normal pressure is 3½-5 psi.

Fig. AC51B—Using Ensign gauge 8276 to set the fuel inlet valve lever to the dimension as indicated at (X) in Fig. AC51.

114C. SYMPTOM. Cold regulator shows moisture and frost after standing.

CAUSE AND CORRECTION. Trouble is due either to leaking valves as per paragraph 114B, or the valve levers are not properly set. For information on setting of valve lever, refer to paragraph 114D.

114D. REGULATOR OVERHAUL. Remove the unit from the engine and completely disassemble, using Fig. AC

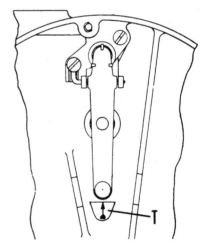

Fig. AC51C—Location of post or boss with stamped arrow for the purpose of setting the fuel inlet valve lever.

51A as a reference. Thoroughly wash all parts and blow out all passages with compressed air. Inspect each part carefully and discard any which are worn.

Before reassembling the unit, note dimension (X—Fig. AC51) which is measured from the face on the high pressure side of the casting to the inside of the groove in the valve lever when valve is held firmly shut as shown in Fig. AC51B. If dimension (X) which can be measured with Ensign gauge No. 8276 or with a depth rule is more or less than ½ inch, bend the lever until this setting is obtained.

A boss or post (T—Fig. AC51C) is machined and marked with an arrow to assist in setting the lever. Be sure to center the lever on the arrow before tightening the screws which retain the valve block. The top of the lever should be flush with the top of the boss or post (T).

NON-DIESEL GOVERNOR

Models B-C-CA-RC

115. Adjust engine speed by shortening or lengthening quadrant lever rod by means of set screw (A—Fig. AC52) to values shown below. Linkage adjustment consists of making sure that the system operates freely and that the link rod connected to the carburetor throttle shaft puts a slight preload on the cross shaft arm when carburetor throttle and hand throttle are in full speed position. This is done by having the link rod 1/32 inch longer than necessary to connect its two attaching levers when they are in the wide open, full speed position. To obtain this effect, it may be necessary to bend the cross shaft arm (12—Fig. AC52A) slightly.

Engine High Idle
  B & C 1850
  CA 2000
Engine Loaded Speed
  B (BE Eng.).............1400
  B & C (CE Eng.).........1500
  CA 1650

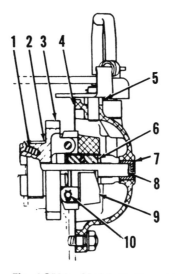

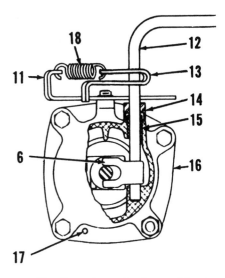

Fig. AC52A—Models B, C, and RC governor assembly. On early production models, the magneto gear hub carries a 10-24 screw (1). Later and current models B and C, and all CA models do not have magneto gear screw (1).

2. Gear bushing
3. Governor gear
4. Housing gasket
5. Control lever bushing
6. Thrust bearing
7. Governor shaft bushing
8. Governor shaft
9. Governor weight
10. Weight pin
11. Manual control lever
12. Cross shaft
13. Spring lever
14. Cross shaft collar
15. Cross shaft bushing
16. Control housing
17. Dowel pin
18. Control lever spring
19. Surge spring

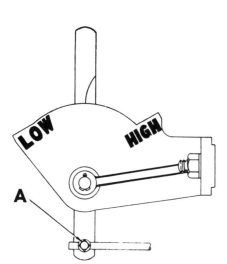

Fig. AC52—Models B, C, CA, and RC governor control lever and quadrant. Maximum engine speed is controlled by lengthening or shortening lever rod by means of set screw (A).

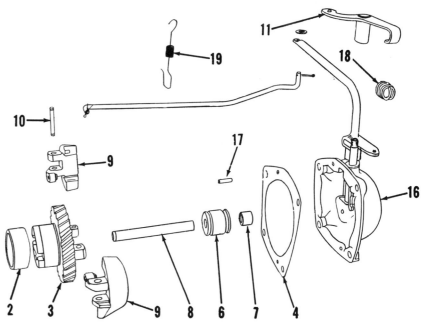

Fig. AC52B—Model CA governor assembly as used on engines equipped with magneto ignition—exploded view. Later and current model B, and later model C governors are similar. Refer to Fig. AC52A for legend.

116. **R&R AND OVERHAUL.** On some tractors, the magneto gear hub carries a 10/24 screw (1—Fig. AC52A) in the magneto gear. On these models to remove the governor unit disconnect the linkage and remove the cap screws which retain the unit to the timing gear cover. The weights unit can now be partially disassembled, including the renewal of thrust bearing (6—Figs. AC52A, AC52B & AC52C). If gear or shaft or shaft bushings require attention, remove magneto unit and screw (1) from gear. The gear, shaft, and remaining portion of weights assembly can now be removed as a unit from the front of timing gear cover.

On all CA models and some tractors of the B-C-RC series the screw (1) is not used. On such models the governor unit can be removed by simply removing the cap screws which retain it to the gear cover.

If shaft or gear is worn, it is advisable to renew it as a single unit to avoid damaging a new shaft by attempting to press the gear on to it.

Bushing (2—Figs. AC52A and AC52B) must be installed with reliefs

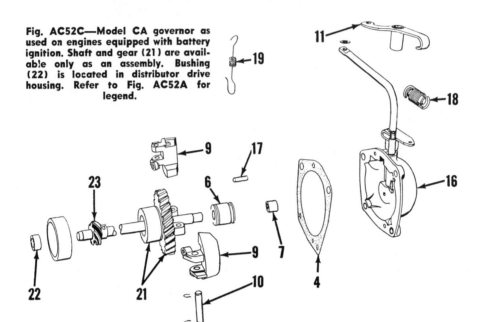

Fig. AC52C—Model CA governor as used on engines equipped with battery ignition. Shaft and gear (21) are available only as an assembly. Bushing (22) is located in distributor drive housing. Refer to Fig. AC52A for legend.

MODEL WD

Fig. AC52D—Model G timing gear end of engine (1) Governor bumper spring adjusting screw, (2) Ignition timing mark, and (3) Carburetor rod.

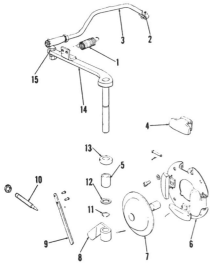

Fig. AC52E—Showing components parts of the Model G governor. Governor weight assembly (6) is accessible after removing timing gear cover.

1. Control spring	9. Bumper spring
3. Carburetor rod	10. Bumper screw
4. Weight	11. Snap ring
5. Roller bearing	12. Washer
6. Weight assembly	13. Seal
7. Thrust cup & shaft	14. Shaft & lever
8. Thrust arm	15. Ball joint

in front face of bushing registered with notches in face of cylinder block. Be sure oil drain hole connecting ignition unit side of bracket to timing gear case is open. When reinstalling the magneto, or battery ignition unit it should be timed as outlined in paragraph 140 or 140A.

Model G

117. **ADJUSTMENT.** With engine running at high idle, turn bumper spring adjusting screw (1—Fig. AC 52D) in until surge is just eliminated. Linkage adjustment consists of making sure that the system operates freely and that carburetor link rod (3) is long enough to put a slight preload on the governor arm when hand

throttle and carburetor are in wide open position. This is done by adjusting rod (3) to a length 1/32 inch longer than required to connect its two attaching levers when the latter are in the wide open position.

Engine High Idle Speed...2100 rpm
Engine Loaded Speed.....1800 rpm

118. **R&R AND OVERHAUL.** To remove the governor unit, it is necessary to first remove the timing gear cover. The weights assembly can then be removed by taking out the four screws which retain it to the cam gear. The governor thrust cup shaft has a .002-.004 clearance in the camshaft. It should be renewed if this clearance exceeds .008 or if cup is worn at the weight contact surface. The weight assembly should be renewed if the hinge pins are worn more than .006. Caution: Check the vent hole located behind number one camshaft journal and drilled to the thrust cup shaft bore. If this hole becomes plugged, the governor will not operate correctly.

The shaft lever located in the timing gear cover can be removed by extracting the snap ring and driving the tapered Groov pin out of the lever.

Models WC-WD-WD45-WF

119. **ADJUSTMENT.** Adjust engine speed by shortening or lengthening control rod (5—Fig. AC52F) by using set screw (4). To adjust linkage follow procedure in paragraph 115.

Engine High Idle
(WC-WF) ....1575 rpm

Engine High Idle
(WD-WD45) ....1720 rpm
Engine Speed Loaded
(WC-WF) ....1300 rpm
Engine Speed Loaded
(WD-WD45) ....1400 rpm

120. **OVERHAUL.** To remove governor weights unit, remove two nuts and cap screw which retain magneto or distributor bracket to engine and withdraw bracket and shaft gear and weights assembly as a single unit. The weights and trust bearing can be renewed at this time.

If shaft, gear, or bushings require attention, the shaft should be removed by bumping out the drive pin (28—Fig. AC52G) on magneto equipped models or remove distributor unit drive gear on battery ignition models. When bumping out the pin, be sure to prevent bending of shaft by bucking up same with a length of pipe introduced through the opening in the bottom of the magneto drive bracket.

On the WD, WC and WF gear (17) is sold only with the shaft assembled, not separately. To avoid damage to the shaft when removing the gear, cool the shaft while heating the gear. Shaft should have .002 running clearance in bushings.

It will be necessary to remove the timing gear cover if the governor throttle shaft and cross shaft are to be overhauled. With cover removed, check cross shaft bushings and, if necessary, renew same. Check governor lever spring eye holes and contact surfaces.

Fig. AC52F—Models WC, WD, and WF engine governor linkage installation. Engine speed adjustment is made at point (4) by increasing or decreasing length of rod (5). Governor linkage on WD and WD45 with battery ignition is similar.

Renew governor lever if the spring holes are elongated and/or contact surfaces are worn flat. Steel washer type shims (24—Fig. AC52H) inserted between governor fork (25) and cross shaft tube (18) control cross shaft end

play. Desired end play is .003-.005.

On the WD, the governor spring (11) is painted red for identification.

When reinstalling the governor, re-time the ignition unit.

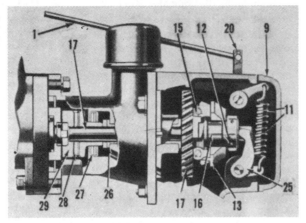

Fig. AC52G—Models WC, WD, and WF governor installation on magneto equipped engines.

1. Carburetor throttle to governor cross shaft linkage
9. Governor cover
11. Governor spring & plunger
13. Weight
15. Weight pivot pin
16. Bearing carrier
17. Governor & magneto drive gear
20. Cross shaft
25. Governor lever
26. Bushing
27. Oil seal
28. Impulse coupling drive retaining pin
29. Impulse coupling

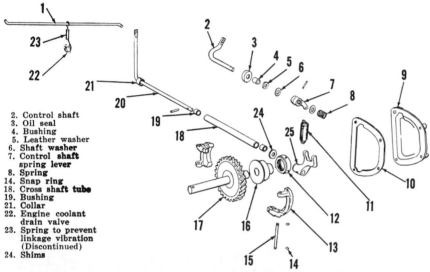

2. Control shaft
3. Oil seal
4. Bushing
5. Leather washer
6. Shaft washer
7. Control shaft spring lever
8. Spring
14. Snap ring
18. Cross shaft tube
19. Bushing
21. Collar
22. Engine coolant drain valve
23. Spring to prevent linkage vibration (Discontinued)
24. Shims

Fig. AC52H—Models WC, WD, and WF engine governor and control linkage as used on magneto equipped engines. Magneto and governor drive gear and shaft (17) are available only as an assembly. Model WD and WD45 equipped with battery ignition are similar except that a distributor drive gear is keyed to the shaft where the magneto drive coupling is located.

Fig. AC52J — Model WD governor to carburetor linkage installation. (1) Cross shaft linkage, (23) Spring to prevent governor surge as installed on earlier models. Current production models are not equipped with this spring. (2) Carburetor choke valve spring. Models WC, WD45 and WF are similar.

NON-DIESEL COOLING SYSTEM

Model G has thermo-syphon cooling; a coolant pump is used on the other models.

RADIATOR & HOSE

All Except Model G

130. Method of removing the radiator is self-evident after examining the tractor. On B, C, and RC, the radiator shell can be removed without disturbing the core and tank unit and the shutters can be removed after removing the shell. On the RC, WC, WD, WD45, and WF, the shell and/or shutters if so equipped can be removed only after removing the radiator assembly from the tractor.

On WC, WD, WD45, and WF, the best method for renewing the lower radiator hose is to remove the water pump.

Model G

131. To R&R radiator, remove fuel tank and fenders as an assembly. Remove engine hood and disconnect the brake return springs from lower side of radiator. Drain radiator and remove upper and lower hose. Remove the two cap screws retaining radiator to clutch housing.

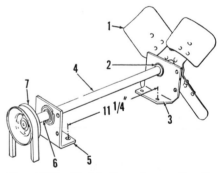

Fig. AC53—Model G engine cooling fan and support assembly.

2. Bearing
3. Support
4. Fan shaft
5. Support
6. Spacer
7. Pulley

FAN

Model G

132. Repairs to fan Fig. AC53 are accomplished by removing the fan, shaft and brackets as a unit. Procedure for doing so is to first remove the hood and hood support. Loosen generator to remove tension from the fan belt. Remove air cleaner and pipe to carburetor. Remove instrument panel from fan shaft bracket and four nuts attaching fan brackets to cylinder head. Fan, pulley and bearings are a

press fit on the shaft and can be disassembled with a suitable puller. Caution: After reinstalling pulley end bearing bracket (5), spacer (6) and pulley (7), press on the fan end bearing bracket (3) and fan blade assembly until a center-to-center measurement of 11¼ inches is obtained between bracket mounting holes.

All Except Model G

133. On B, C, CA, and RC models, to R&R the fan blades assembly, it is necessary to first remove the radiator.

134. On WC, WD, WD45, and WF equipped with ball bearing type water pumps (ball type has no lubrication fitting), the R&R of fan blades necessitates a special procedure as follows: Unbolt fan from pump hub and drop fan to bottom of radiator shell. Disconnect lower hose and detach pump from engine. Fan unit can now be removed from bottom of radiator shell.

On WC, and WF models equipped with plain bearing water pumps (provided with a lubrication fitting), the

fan blades unit can be removed from the tractor without disturbing the pump.

WATER PUMP

135. **R&R PUMP.** On B, C, CA, and RC, to R&R the pump, it is necessary to first remove the radiator shell and the radiator.

On WC, WD, WD45, and WF, to R&R pump, remove hood, fan belt, and fan retaining cap screws. Allow fan to rest at bottom of shell. Disconnect radiator hose. Remove pump to engine cap screws and lift pump up and out towards left side of tractor. Remove fan from bottom of shell.

136. **OVERHAUL B-C-CA-RC.** To disassemble pump, remove fan blades, pulley, and rear cover. Attach a suitable puller to tapped holes in impeller (13—Fig. AC53A), and pull impeller from shaft (5). Carbon thrust washer (3) and seal assembly (14) can be removed from impeller after removing snap ring (4) from impeller. Shaft and bearing assembly (5) and fan hub flange (7) can be pressed out of front of pump body after removing bearing lockwire (9). Shaft can be pressed out of fan hub.

Surface of pump body contacted by the carbon thrust washer must be smooth and true. When pressing impeller on shaft, do not collapse seal; allow approximately 1/16 inch at "A". Fan hub flange should be pressed on

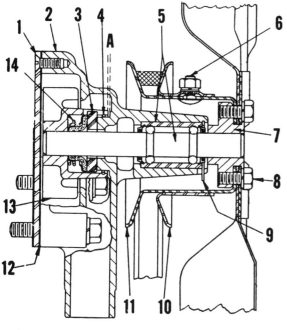

Fig. AC53A — Models B, C, CA, and RC water pump assembly. When pressing impeller on pump shaft, allow 1/16 inch impeller clearance as shown at (A).

A. Impeller Clearance
1. Rear cover
2. Pump body
3. Carbon thrust washer
4. Seal retaining snap ring
5. Pump shaft and bearing assembly
6. Pulley lock bolt
7. Fan hub flange
8. Fan retaining cap screw
9. Bearing lock wire
10. Pulley flange (front)
11. Pulley flange (rear)
12. Rear cover gasket
13. Impeller
14. Seal assembly

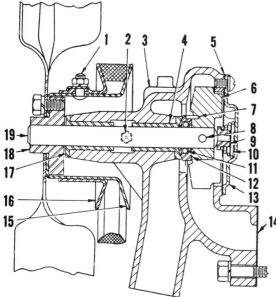

Fig. AC53B—Early production models WC, and WF were equipped with a plain bushing water pump. Bushing type pumps can be identified by an external lubrication fitting for bushings (4 & 17).

1. Pulley lock bolt
4. Rear bushing
5. Rear cover gasket
6. Impeller
7. Thrust washer driving pin
8. Impeller pin
9. Spring retainer
10. Thrust spring
11. Shaft seal
12. Thrust washer
14. Flange gasket
15. Pulley flange (rear)
16. Pulley flange (front)
17. Front bushing

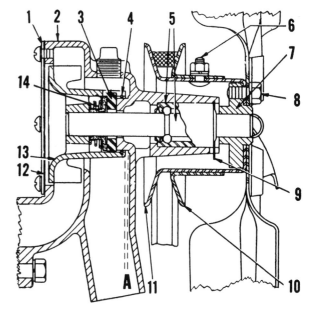

Fig. AC53C — Later production models WC and WF, and all WD and WD45 engines are equipped with a ball bearing type pump. When pressing impeller on pump shaft, allow 1/16 inch impeller clearance as shown at (A).

A. 1/16 inch impeller clearance
1. Rear cover
2. Pump body
3. Carbon thrust washer
4. Seal snap ring
5. Pump shaft & bearing
6. Pulley lock bolt
7. Fan hub flange
8. Bearing snap ring
9. Pulley flange (front)
10. Pulley flange (rear)
11. Rear cover gasket
12. Rear cover gasket
13. Impeller
14. Seal assembly

shaft to the position that will align the fan pulley with the crankshaft pulley.

137. OVERHAUL WC-WF BUSHING TYPE. To disassemble pump, remove nut, fan hub, and key. If pump does not have nut on front of shaft, use a press or puller to remove fan hub. Remove cover screws and cover (13—Fig. AC53B). Register impeller vanes with notches in pump body and withdraw shaft (19) and impeller (6)

from rear of pump body (3). Thrust washer (12) and seal (11) can be removed from shaft. To remove impeller from shaft, remove pin (8) and use a puller or press. Rear surface of rear bushing (4) contacted by the carbon thrust washer (12) must be smooth and true. When pressing bushings into pump body, use a mandrel .001 inch larger than shaft diameter to prevent crushing bushings. Be sure notch in edge of thrust washer is engaged with

driving pin (7) in impeller and retainer (9) and spring (10) are installed correctly between end of shaft and cover. Fan hub flange should be pressed on shaft to the position that will align the fan pulley with the crankshaft pulley.

138. WC - WD - WD45 - WF BALL BEARING TYPE. On these pumps shown in Fig. AC53C, follow same procedure as outlined for models B, C, CA, and RC in paragraph 136.

NON-DIESEL ELECTRICAL SYSTEM

ELECTRICAL EQUIPMENT

139. DISTRIBUTORS. Model CA is equipped with a Delco-Remy 1111735 distributor. The model G distributor is a Delco-Remy 1111708. Models WD and WD45 are equipped with a Delco-Remy 1111745 distributor. Specifications are as follows:

1111708
Rotation, drive end............. CC
Advance data, distributor
degrees and rpm
  Start ....0-2 @ 400
  Intermediate ....2-4 @ 600
  Maximum ....5-7 @ 900
Contact gap 0.022
Cam Angle, degrees...........25-34

1111735 & 1111745
Rotation, drive end............. CC
Advance data, distributor
degrees and rpm
  Start .... 2-5 @ 225
  Intermediate .... 3-6 @ 300
  Intermediate .... 6-9 @ 375
  Maximum ....14-16 @ 550
Contact gap 0.022
Cam Angle, degrees...........25-34

GENERATORS. All valve-in-head models used either a Delco-Remy 1101357 or a 1101413 generator and most of them obtained regulation with a 3 step resistance switch Delco-Remy 479G. The model G generator is a Delco-Remy 1101854. Specifications are as follows:

1101357 & 1101413
Brush spring tension...........19 oz.
Field draw, volts............... 6.0

Field draw, amperes........ 3.5- 4.5
Hot output, amperes 10.0-13.0
Hot output, volts 7.4- 7.7
Hot output, rpm................2200

1101854
Brush spring tension..........19 oz.
Field draw, volts............... 6.0
Field draw, amperes........ 2.0-2.3
Hot output, amperes........ 7.5-9.5
Hot output, volts........... 7.2-7.4
Hot output, rpm...............3000

STARTING MOTORS. Prior to about 1949 the valve-in-head models were equipped with a Delco-Remy starter either 1107017 or 1107043. Later series valve-in-head models used 1107951, except the CA which has a 1107096 Delco-Remy starter. Model G has a 1109605 starter. Specifications are as follows:

1107017 & 1107043 & 1107096
Volts 6
Brush spring tension........26-28 oz.
No load test, volts............. 5.65
No load test, amperes........... 70
No load test, rpm............. 5500
Lock test, volts............... 3.25
Lock test, amperes 550
Lock test, torque, ft.-lbs........ 11

1107951
Volts 6
Brush spring tension........24-28 oz.
No load test, volts............. 5.0
No load test, amperes........... 60
No load test, rpm............. 6000
Lock test, volts............... 3.0
Lock test, amperes 600
Lock test, torque, ft.-lbs........ 15

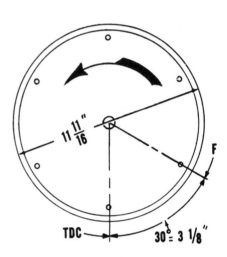

Fig. AC53D—Models WC, WF and early WD flywheel viewed from rear. "F" or "Fire" is the only mark which appears on the flywheel when viewed through inspection port in lower side of clutch housing. Later WD and all WD45 flywheels are similar and are marked with a "DC" mark.

IGNITION TIMING

Models B-C-CA-RC-WC-WD (Magneto)

140. Fairbanks-Morse model FMJ or FMX-4B3 magneto of clockwise rotation (viewed from impulse coupling) and a 30 degree lag angle impulse coupling is used on these engines.

Set breaker contacts to .020 gap. Magneto is timed to the advance timing mark "F", or "Fire" which is 30 deg. BTC or 3⅛ inches BTC, on WD flywheel. To time the magneto to the engine, first crank engine until No. 1 piston is on its compression stroke and the flywheel timing mark is centered in inspection port (located in bottom of clutch housing).

Fig. AC53E — Model CA equipped with battery ignition unit. Distributor receives its drive from governor gear and shaft.

TG. Timing gear cover
11. Manual throttle lever
12. Cross shaft
16. Control housing
18. Control lever spring
31. Distributor drive housing

This method sets the fully advanced or running timing of the distributor. If a timing light is available the timing should be rechecked by running the engine at fast idle (above 1100 rpm) at which time the flash should occur when flywheel mark "FIRE" or "F" is centered in the inspection port.

Static timing is top center which is indicated on the flywheel by the mark "DC".

Remove magneto distributor cover and rotate impulse coupling backward until the distributor rotor is aligned with the timing lug (projection in upper left corner of distributor compartment). With the rotor held in this position, (breaker contacts just opening) install magneto to engine. Elongated slots in magneto mounting flange will permit a closer distributor rotor-to-lug alignment if same has been disturbed during installation. If magneto coupling cannot be engaged without turning magneto shaft more than 10 degrees (as might be the case where the magneto is being retimed after overhauling the governor shaft), the drive gear must be remeshed to the camshaft gear. Connect No. 1 spark plug wire to the No. 1 distributor terminal (directly over rotor and lug). Firing order is 1-2-4-3.

Models CA-WD-WD45 (Battery Ignition)

140A. Crank engine until No. 1 piston is coming up on compression stroke then crank slowly and stop when flywheel mark "Fire" is in center of inspection port (left side of torque tube on CA; bottom of clutch housing on WD and WD45.) Place distributor rotor in position where metal strip on top of same is pointing toward battery lead terminal on distributor body. With contacts set to .022 gap, place distributor in mounting bracket and engage the driving members. Now rotate distributor cam as far as it will go in direction of its normal rotation and while holding it in this position, rotate the distributor body until the contacts just start to open. Now tighten the mounting bracket to lock the distributor in this position.

Model G

140B. Delco-Remy model 1111708 distributor, of counter clockwise rotation (viewed from gear end) is used on this tractor. Set breaker points to .020 gap. To time the distributor to the engine, crank the engine until No. 1 piston (timing gear end) is on its compression stroke and the notched timing mark located on crankshaft pulley is centered in inspection hole, as shown at (T), in Fig. AC53G. Notched timing mark is located 3 degrees before TC.

Install the distributor unit to the cylinder block opening with the rotor (finger) under the number one terminal in the cap and with the primary wire terminal pointing at 2 o'clock approximately. Rotate distributor housing anti-clockwise until breaker contacts have just opened, then tighten distributor clamp. Firing order is 1-3-4-2.

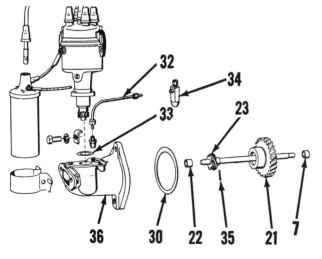

Fig. AC53F—Model CA battery ignition unit and drive — exploded view. Shaft and gear (21) are available only as an assembly.

7. Governor shaft bushing
21. Governor shaft & gear
22. Governor shaft bushing
23. Distributor drive gear
30. Gasket
32. Oil tube
33. Gasket
34. Oil tube tee
35. Retaining pin
36. Drive housing

Fig. AC53G — Model G ignition timing mark, located on face of crankshaft pulley, is viewed through inspection port in hood support as shown.

DIESEL ENGINE AND COMPONENTS

Fig. AC54—View showing left side of WD45 Diesel engine. The engine is fitted with a Bosch PSB injection pump and pintle nozzles.

Fig. AC54A—View showing right side of engine. Removal of Nos. 2, 3, 4 or 5 energy cells requires removal of the air intake manifold.

R&R ENGINE WITH CLUTCH

141. To remove the engine and clutch as a unit, first drain the cooling system, and if engine is to be disassembled, drain the oil pan. Remove the hood and radiator. Disconnect the fuel line running from primary filter to transfer pump. Disconnect the return line from injection pump and leak off line to top of fuel tank. Disconnect throttle rod, generator wires and temperature indicating bulb. Remove universal joint retaining pin from steering worm shaft and slide steering shaft and universal joint rearward.

141A. Support engine in a suitable hoist as shown in Fig. AC54B, and unbolt side channels from front support. Slide front support forward until front cap screws in side channels can be reinstalled in rear holes of front support. If tractor has an adjustable front axle, the nut and lock washer must be removed from the radius rod pivot bolt. If desired for convenience, the radius rod support may be removed by removing the three attaching bolts.

Unbolt and remove the engine front support from side channels.

141B. Remove cap screws retaining engine to clutch housing and slide engine forward until free of clutch housing dowel pins and clutch shaft. Raise front of engine and at the same time slide the engine forward until the clutch clears the clutch housing. Lift engine from tractor.

CYLINDER HEAD

142. To remove the cylinder head, first drain cooling system and remove hood. Remove the oil line which runs from the oil gallery to the cylinder head. Disconnect heat indicator bulb from water outlet casting on top of head and disconnect water pump from front of head. Disconnect by-pass hose from thermostat housing, remove cap screw retaining the water inlet pipe to cylinder head and lay by-pass line and inlet pipe assembly forward and out of the way. Remove thermostat housing from water outlet casting, and water outlet casting from head. Disconnect heater cable from manifold and remove the manifold air inlet tube. Disconnect the diesel fuel sys-

tem supply and return lines, disconnect the intermediate and secondary (or final) fuel filters mounting bracket and lay filters and bracket over and out of the way. Disconnect nozzle lines from injection pump and nozzles. Remove valve cover, rocker arms assembly and push rods. Remove the cylinder head retaining stud nuts and lift cylinder head from tractor. If desired for convenience, the intake and exhaust manifolds may be removed. If Nos. 2, 3, 4 or 5 energy cells are to be removed it will be necessary to remove the manifolds.

When installing the cylinder head, tighten the stud nuts progressively, from center outward and to a torque value of 95-100 Ft.-Lbs.

Fig. AC54B—Tractor with hood, radiator and engine front cover (timing cover) removed.

VALVES AND SEATS

142A. Intake and exhaust valves are not interchangeable and the intake valves on all models seat directly in cylinder head with a face and seat angle of 45 degrees. The exhaust valves seat on renewable ring type inserts with a face and seat angle of 45 degrees. The method of removing the renewable type valve seat inserts is shown in Fig. AC55. Valve seat width is $\frac{3}{32}$-inch for both the intake and the exhaust. Seats can be narrowed, using 15 and 75 degree stones. Intake and exhaust valves have a stem diameter of 0.3095-0.3100.

Adjust the intake valve tappet gap to 0.010 hot and the exhaust valve tappet gap to 0.019 hot.

VALVE GUIDES AND SPRINGS

142B. Guides can be pressed or driven from cylinder head if renewal is required. Press guides into cylinder head with smaller O.D. of guide up, until top of the guide is $\frac{5}{16}$-inch above the valve cover gasket surface of cylinder head. Maximum allowable valve stem to guide clearance is 0.006. Ream new guides after installation, to provide the recommended stem to guide clearance of 0.0015-0.003. Refer to paragraph 142A for valve stem diameter.

142C. Intake and exhaust valve springs are interchangeable in all models. Renew any spring which is rusted, discolored or does not meet the load test specifications which follow:

Spring free length.........2$\frac{1}{32}$ inches
Renew if less than.........1$\frac{29}{32}$ inches
Pounds pressure
  at 1$\frac{3}{4}$ inches................ 40- 50
Pounds pressure
  at 1$\frac{13}{32}$ inches................105-115

Fig. AC55—Method of removing a valve seat insert.

VALVE TAPPETS
(CAM FOLLOWERS)

142D. The 0.560-0.561 diameter mushroom type tappets operate directly in the unbushed cylinder block bores with a suggested clearance of 0.0005-0.0025. Maximum allowable clearance is 0.0035.

To remove the tappets, it is first necessary to remove the camshaft as outlined in paragraph 144. Tappets are available in standard size only.

ROCKER ARMS

143. The hollow rocker arm shaft is drilled for lubrication to each rocker arm bushing. Lubricating oil to the drilled cylinder head passage and slotted oil stud (S—Fig. AC55A) is supplied by an external oil line which is connected to the main oil gallery on left side of engine. CAUTION: Always check the slotted stud to be sure the slot is facing the oil passage in the head. If oil does not flow from the hole in the top of each rocker arm, check for foreign material in the external oil line or in the cylinder head passage.

Fig. AC55A—Rocker arms assembly is lubricated via the slotted stud (S).

Fig. AC55B—Rocker arms are fitted with renewable type valve stem contact buttons which can be removed after extracting the retaining snap rings as shown.

The procedure for disassembling and reassembling the rocker arms from the shaft is evident. Check the rocker arm shaft and the bushing in each rocker arm for excessive wear. Maximum allowable clearance between the shaft and bushings is 0.005. When installing new bushings, make certain that oil hole in bushing is in register with oil hole in rocker arm and ream the bushings to provide a clearance of 0.001-0.002 for the 0.8405-0.841 diameter rocker arm shaft. When installing the rocker arm shaft, make certain that the oil metering holes in the shaft point toward push rods instead of valve stems.

Inspect the valve stem contact button in the end of each rocker arm for being mutilated or excessively loose. If either condition is found, renew the contact button. Extract the button retaining snap ring as shown in Fig. AC55B and remove the button and oil wick. Install new oil wick and button and test the button for a free fit in the rocker arm socket.

Note: If a new contact button has any binding tendency in the rocker arm socket, use a fine lapping compound and hand lap the mating surfaces.

VALVE TIMING

143A. Valves are properly timed when timing marks on camshaft gear and crankshaft gear are in register. The single punch marked tooth space on the camshaft gear should be meshed with the single punch marked tooth on the crankshaft gear, as shown at (X) in Fig. AC56.

CRANKCASE FRONT COVER

143B. To remove the crankcase front cover (timing gear cover), first drain the cooling system and remove the

radiator. Remove retaining pin from steering shaft universal joint and slide steering shaft rearward. Support tractor and slide front support forward as described in paragraph 141A. Remove fan belt and crankshaft pulley. Unbolt engine front support (S — Fig. AC56B) from side channels, loosen clamp bolt (CB) and remove bracket (B).

Remove the cap screws from the front of the timing gear cover and the three retaining cap screws from the injection pump mounting flange (injection pump will stay in place) and withdraw timing gear cover from front of engine.

The crankshaft front oil seal (8—Fig. AC56A) which is retained in the crankcase front cover can be renewed at this time. The spring loaded oil seal should be installed with lip of same facing inward toward timing gears.

Reinstall the crankcase front cover by reversing the removal procedure and make certain that copper washers (4) are installed on the bottom three cover retaining cap screws.

TIMING GEARS

143C. The camshaft gear and/or crankshaft gear can be removed without removing their respective shafts from the engine. To remove the gears, it is necessary to use a suitable puller after the crankcase front cover has been removed as outlined in paragraph 143B.

Before removing either of the gears, check the gear backlash which should be 0.001-0.003. If the backlash exceeds 0.007, either one or both gears

Fig. AC56B — Front support (S) and bracket (B) can be removed together.

should be renewed. Before removing the camshaft gear and after removing the camshaft thrust plate retaining cap screws as shown in Fig. AC57, check the camshaft end play by inserting a feeler gage between the rear face of the camshaft thrust plate and the front face of the camshaft front bearing journal, as shown in Fig. AC-57B. The thickness of the feeler gage that can be inserted represents the camshaft end play which should be 0.003-0.008. If the camshaft end play is excessive (more than 0.014), and a new thrust plate will not correct the condition, it will be necessary to renew the camshaft gear and recheck the end play; or, if the same gear is to be used, it will be necessary to file the required amount from the rear face of the gear hub as shown in Fig. AC57A. For example, if the measured end play is 0.015; file at least 0.007 from the gear hub so as to bring the end clearance within the suggested limits.

Timing gears are marked with a letter "S" if the gears are standard size; or, the gears are marked with a number within either an "O" or a letter "U". The letter "O" indicates an oversize gear and the letter "U" indicates

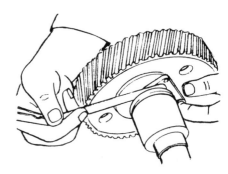

Fig. AC57B — Checking the camshaft end play. The amount of end play is equal to the thickness of the feeler gage that can be inserted between the shaft journal and the thrust plate.

Fig. AC56—The single punch marked tooth space on the camshaft gear should be meshed with the single punch marked tooth on the crankshaft gear as shown at (X).

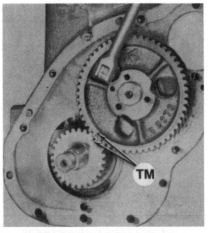

Fig. AC57—Removing the camshaft thrust plate retaining cap screws.

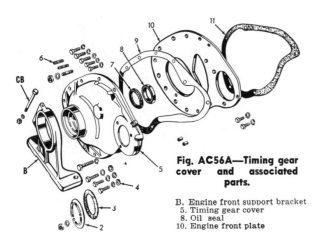

Fig. AC56A—Timing gear cover and associated parts.

B. Engine front support bracket
5. Timing gear cover
8. Oil seal
10. Engine front plate

Fig. AC57A — Excessive camshaft end play can be corrected by filing the required amount of metal from rear face of the camshaft gear hub. Refer to text.

an undersize gear. The enclosed number gives the deviation from a standard size gear in thousandths of an inch. When installing new gears, always use the same size gears as were removed and check the backlash to make certain that the value is within the clearance limits. Due allowance should be made, however, for wear on the other gear if only one gear is renewed.

Usually, new gears are punched with timing marks before they leave the factory. If, however, a new gear is not so marked, it is important to transfer the marks from the old gear to the new. To do so, proceed as follows: Place the old gear on top of the new one and using a piece of keystock, align the key ways of both gears. Place a straight edge against the gear teeth and in line with the timing mark on the old gear. Locate and punch mark the appropriate tooth (or tooth space) on the new gear. Refer to Fig. AC58.

To facilitate installation of a new crankshaft gear, boil the gear in oil for a period of 15 minutes prior to installation. When installing the camshaft gear, mesh the single punch marked camshaft gear tooth space with the single punch marked crankshaft gear tooth as shown in Fig. AC56.

NOTE: When installing the camshaft gear with the camshaft in the engine, remove the oil pan and place a heavy bar against the side of a cam to hold the camshaft in a forward position while the gear is being drifted on. This procedure will eliminate the possibility of loosening the soft plug which is located in the cylinder block at the rear of the camshaft.

CAMSHAFT AND BEARINGS

The camshaft is supported in four precision steel-backed, babbitt-lined bearings. The shaft journals have a normal operating clearance of 0.002-0.0046 in the bearings· If the journal clearance exceeds 0.014 the bearings and/or shaft should be renewed. To renew the bearings, follow the procedure outlined in paragraph 144A. To renew the camshaft, refer to the following paragraph.

144. **CAMSHAFT.** To remove the camshaft, first remove the crankcase front cover as outlined in paragraph 143B, then, proceed as follows: Remove valve cover, rocker arms assembly and push rods. Insert long wooden dowel pins down through the push rod openings in crankcase and tap the dowels into the tappets (cam followers). Lift dowels up, thereby holding tappets away from camshaft and hold dowels in the raised position with spring type clothes pins. Re-

move oil pan and oil pump. Working through openings in camshaft gear, remove the camshaft thrust plate retaining cap screws and withdraw camshaft from engine.

Recommended camshaft end play should be 0.003-0.008. Refer to timing gears (paragraph 143C) for the method of checking and adjusting the camshaft end play.

Check the camshaft against the values which follow:

No. 1 (front) journal
  diameter 1.998-1.999
No. 2 journal diameter.....1.998-1.999
No. 3 journal diameter.....1.998-1.999
No. 4 (rear) journal
  diameter 1.248-1.249

When installing the camshaft, reverse the removal procedure and make certain that the valve timing marks are in register as shown in Fig. AC56. Caution: Make certain that the drilled camshaft thrust plate retaining cap screw is installed in the lower hole.

144A. **CAMSHAFT BEARINGS.** To remove the camshaft bearings, first remove the engine as in paragraph 141; then, remove camshaft as in the preceding paragraph 144. Remove clutch, flywheel and the engine rear end plate. Extract the soft plug from behind the camshaft rear bearing and remove the bearings.

An approved method for removing the bearings is to split the bearings with a hack saw blade and drive the bearings out with a chisel. Exercise special care during this operation, however, to avoid damaging the crankcase bearing bores. Clean the camshaft bearing oil holes in crankcase.

Using a closely piloted arbor, install the bearings so that oil hole in bearings is in register with oil holes in the crankcase.

The inside diameter of the camshaft bearings after installation should be as follows:
No. 1 (front) 2.0010-2.0026
No. 2 ....2.0010-2.0026
No. 3 ....2.0010-2.0026
No. 4 (rear) 1.2510-1.2526

Although the camshaft bearings are presized, it is highly recommended that the bearings be checked after installation for localized high spots. The camshaft bearing journals should have a normal operating clearance in the bearings of 0.002-0.0046.

When installing the soft plug at rear camshaft bearing, use Permatex or equivalent to obtain a better seal.

CONNECTING ROD AND PISTON UNITS

144B. Piston and connecting rod units are removed from above after removing cylinder head and oil pan. The procedure for removing the cylinder head is outlined in paragraph 142.

Cylinder numbers are stamped on the connecting rod and the cap. When reinstalling the rod and piston units, make certain the cylinder identifying numbers are in register and face opposite the camshaft side of engine.

Tighten the connecting rod bolts to a torque of 30 to 40 ft. lbs.

PISTONS, RINGS AND SLEEVES

144C. Each piston is fitted with five rings; three ⅛-inch wide compression rings and two $\frac{3}{16}$-inch wide oil con-

Fig. AC58—In rare cases, replacement timing gears are not punched with timing marks. When such cases are encountered, transfer timing marks from the old gear to the new, using a straight edge and keystock as shown.

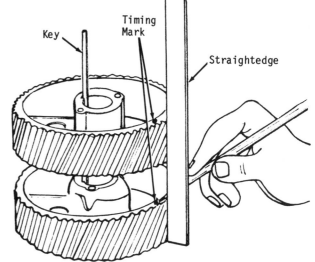

trol rings. Check the pistons and rings against the values of which follow:

Ring Groove Width:

Top ....0.127 -0.128
Second ....0.126 -0.127
Third ....0.126 -0.127
Fourth 0.1880-0.1895
Fifth 0.1880-0.1895

Piston Ring Side Clearance:

Top ....Desired 0.003-0.005
     (Maximum 0.007)
Second ....Desired 0.002-0.004
     (Maximum 0.0055)
Third ....Desired 0.002-0.004
     (Maximum 0.0055)
Fourth ....Desired 0.0015-0.0035
     (Maximum 0.005)
Fifth ....Desired 0.0015-0.0035
     (Maximum 0.005)

End Gap, All Rings........0.009-0.014

Sleeves are of the renewable, wet type and should be renewed if out of round exceeds 0.003 and/or taper exceeds 0.009.

The desired clearance between piston skirt and sleeve is 0.004-0.005. The maximum allowable clearance is 0.0095.

Piston to sleeve clearance can be considered satisfactory when a spring pull of 2 to 5 lbs. is required to withdraw a 0.002 thick, ½-inch wide feeler gage.

R&R CYLINDER SLEEVES

145. The wet type cylinder sleeves can be renewed after removing the connecting rod and piston units. Refer to paragraph 144B. Coolant leakage at bottom of sleeves is prevented by two rubber "O" rings. The cylinder block is counter-bored at the top to receive the sleeve flange and the head gasket forms the coolant seal at this point. The sleeves can be removed, using a special puller; or, the sleeves can be removed by placing a wood block against bottom edge of sleeve and tapping the block with a hammer.

Before installing new sleeves, thoroughly clean the cylinder block, paying particular attention to the seal seating surfaces at bottom and the counterbore at top. All sleeves should enter crankcase bores full depth and should be free to rotate by hand when tried in bores without "O" rings. After making a trial installation without "O" rings, remove the sleeves and install the "O" rings making certain that the "O" rings are not twisted. To facilitate installation of the sleeves and to keep from cramping or causing the "O" rings to bulge, coat the "O" ring with hydraulic brake fluid

or a thick soap solution. Sleeve standout should be 0.004-0.006 above cylinder block after installation.

PISTON PINS

145A. The full floating type 0.99955-0.99975 diameter piston pins are retained in the piston pin bosses by snap rings and are available in standard size only. Check the piston pin fit against the values which follow: Piston pin clearance in piston 0.00025T-0.00045L. Piston pin clearance in rod bushing 0.00015-0.00085.

CONNECTING RODS AND BEARINGS

145B. Connecting rod bearings are of the renewable, steel-backed, copper-lead-lined, slip-in type. The bearings can be renewed after removing oil pan and bearing caps. When installing new bearing shells, make certain that the bearing shell projections engage the milled slot in connecting rod and bearing cap and that cylinder numbers on the rod and cap are in register and face opposite to camshaft side of engine. Bearing inserts are available in standard size as well as undersizes of 0.010, 0.020 and 0.040. Check the crankshaft crankpins and the bearing inserts against the values which follow:

Crankpin diameter
  (Standard) 1.9975–1.9985
Rod bearing running
  clearance (desired) ...0.0015–0.0035
     (Maximum 0.006)

Rod side play (desired) ..0.002 –0.007
     (Maximum 0.013)
Rod bolt torque
  (Ft.-Lbs.) 30–40

CRANKSHAFT AND MAIN BEARINGS

145C. The crankshaft is supported in seven steel-backed, copper-lead-lined, slip-in precision type main bearings. Main bearings can be renewed after removing oil pan and main bearing caps. Desired crankshaft end play is 0.002-0.007. Maximum allowable end play is 0.011. Crankshaft end play is corrected by renewing the center main bearing shells.

To remove the crankshaft, first remove the engine as outlined in paragraph 141. Remove clutch, flywheel and engine rear end plate. Remove valve cover, rocker arms assembly and push rods. Remove oil pan, oil pump, and rod and main bearing caps. Remove crankcase front cover (timing gear cover), unbolt camshaft thrust plate, withdraw camshaft and remove the engine front end plate. Lift crankshaft from engine.

Check crankshaft and main bearings against the values which follow:
Crankpin diameter
  (Standard) 1.9975-1.9985
Main journal diameter
  (Standard) ....2.497 -2.498
Main bearing running
  clearance (desired) ...0.0023-0.0045
     (Maximum 0.011)
Main bearing bolt torque
  (Ft.-Lbs.) 125-135

Three-quarter view of WD45 Diesel engine.

Main bearings are available in standard size as well as undersizes of 0.010, 0.020 and 0.040.

CRANKSHAFT REAR OIL SEAL

145D. The crankshaft rear oil seal is of the one piece spring loaded type. The seal is retained in the oil seal retainer plate (15—Fig. AC59) which is retained to the front face of the engine rear end plate by six screws. Circular seal (13) fits around seal retainer (15) and forms the rear seal for the oil pan. To remove the spring loaded oil seal (14), first remove the flywheel as outlined in paragraph 146A and pry the seal out of the groove in the seal retainer plate. To remove the circular seal (13), it is necessary to support rear of engine and remove the engine rear end plate.

Fig. AC59A — The oil pump drive gear is retained to the drive shaft by a pin. After grinding off the end of the pin, the pin can be drifted out as shown.

OIL PAN

146. The procedure for removing and reinstalling the oil pan is evident. The following, however, should be observed.

Five of the oil pan retaining cap screws are longer than the others. These longer cap screws pass through the crankcase front cover (timing gear cover) and into the front face of the oil pan. The bottom three of these cap screws must be fitted with copper washers. When installing the oil pan, install all of the cap screws loosely; then, tighten the front five cap screws before tightening any of the others.

FLYWHEEL

146A. To remove the flywheel, first remove the engine as outlined in paragraph 141. Then, remove the clutch. The flywheel is held to the crankshaft flange by cap screws and is doweled for alignment. Always check the timing marks of the old and new flywheel when renewing. Examine the oil seal flange on the front of the flywheel for nicks and burrs. If any are found, smooth with a fine stone

or crocus cloth. Use caution when reinstalling to prevent injury to the oil seal. Torque the flywheel retaining cap screws to 95-105 Ft.-Lbs.

The starter ring gear can be removed by drilling and splitting same with a cold chisel. To install a new ring gear heat to 300° F.-400° F. by boiling in oil or heating evenly with a torch. Install new ring gear with beveled edge of teeth facing rearward. Note: Do not overheat as ring gear may be annealed.

OIL PUMP

146B. The oil pump, which is gear driven from the camshaft, can be removed after removing the oil pan.

To disassemble the pump, proceed as follows: File off the head of the pin which retains the pump driving gear to the pump drive shaft and using a small punch as shown in Fig. AC59A, remove the pin. Remove cover from pump housing and withdraw drive shaft and gear and idler gear and shaft from pump housing. To remove the driver gear from the drive

shaft, press gear further up the shaft until the snap ring is exposed. Remove the snap ring and press the gear off the shaft. Inspect the oil seal (9— Fig. AC60) in the pump body. There is no need to remove the seal unless the seal is damaged. If however, the seal is removed, always install a new seal.

Pumping gears should not have over 0.020 backlash nor more than 0.006 end play in pump body. DO NOT in-

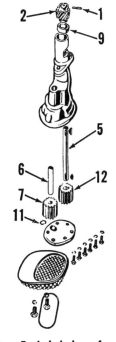

Fig. AC59 — Chankshaft rear oil seal (14) is of the spring loaded type and is carried in seal retainer (15) which is bolted to the engine rear end plate (12). Circular seal (13) forms the rear seal for the oil pan.

Fig. AC60 — Exploded view of early diesel engine oil pump. Later models are similar, but differ in details.

1. Pin	7. Idler gear
2. Gear	9. Oil seal
5. Drive shaft	11. Snap ring
6. Idler shaft	12. Drive gear

crease spring pressure on oil pressure relief valve as a cure for a worn pump.

To prevent possible damage to the drive shaft oil seal, install the pump drive shaft oil seal, then install the pump drive shaft as follows: With the oil seal in place, temporarily install the idler shaft in the pump body at oil seal end of pump. With the idler shaft extending through the oil seal, insert the drive shaft into the pump body, from drive end, until drive shaft butts against the idler shaft. Press drive shaft into position, forcing the idler shaft out of the pump body. The remainder of the assembly procedure is evident.

OIL PRESSURE RELIEF VALVE

146C. Normal oil pressure of 25 psi is controlled by a spring-loaded oil pressure relief valve. The pressure is adjusted by a slotted screw (26—Fig. AC60A) on left front side of crankcase. If difficulty is encountered

when attempting to regulate the oil pressure, check the drilled cap screw which holds the camshaft gear thrust plate in place. If the screw is too long, it will interfere with the valve action. The drilled cap screw dimensions are $\frac{7}{16}$-14 x $\frac{9}{16}$ inch.

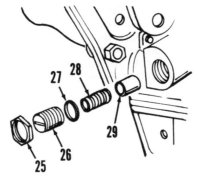

Fig. AC60A—Oil pressure relief valve exploded from left front side of cylinder block.

25. Nut
26. Pressure adjusting screw
27. Gasket
28. Relief valve spring
29. Relief valve

147. **QUICK CHECKS—UNITS ON TRACTOR.** If the diesel engine does not start or does not run properly, and the diesel fuel system is suspected as the source of trouble, refer to the Diesel System Trouble Shooting Chart and locate points which require further checking. Many of the chart items are self-explanatory; however, if the difficulty points to the fuel filters, injection nozzles and/or injection pump, refer to the appropriate paragraphs which follow:

FUEL FILTERS

The fuel filtering system consists of a metal sediment bowl, primary filter of the replaceable element type, an intermediate filter of the replaceable element type, and a final (secondary) filter of the sealed type.

147A. **CIRCUIT DESCRIPTION AND MAINTENANCE.** Fuel from the fuel tank flows through a metal sediment bowl (A—Fig. AC60B) which should be removed, drained and cleaned each

Fig. AC60B—Metal sediment bowl (A) and the first stage, renewable element type fuel filter (B).

DIESEL FUEL SYSTEM

The diesel fuel system consists of three basic units; the fuel filters, injection pump and injection nozzles. When servicing any unit associated with the fuel system, the maintenance of absolute cleanliness is of utmost importance. Of equal importance is the avoidance of nicks or burrs on any of the working parts.

Probably the most important precaution that service personnel can impart to owners

of diesel powered tractors, is to urge them to use an approved fuel that is absolutely clean and free from foreign material. Extra precaution should be taken to make certain that no water enters the fuel storage tanks. This last precaution is based on the fact that all diesel fuels contain some sulphur. When water is mixed with sulphur, sulphuric acid is formed and the acid will quickly erode the closely fitting parts of the injection pump and nozzles.

DIESEL SYSTEM TROUBLE SHOOTING CHART

	Sudden Stopping of Engine	Lack of Power	Engine Hard to Start	Irregular Engine Operation	Engine Knocks	Engine Smoking	Excessive Fuel Consumption
Lack of fuel	★	★	★	★			
Water or dirt in fuel	★	★	★	★			
Clogged fuel lines	★	★	★	★			
Inferior fuel	★	★	★	★			★
Faulty transfer pump	★	★	★	★			
Faulty injection pump timing		★	★	★	★	★	★
Air traps in system	★	★	★	★			
Clogged fuel filters		★	★	★			
Deteriorated fuel lines	★						★
Air leak in suction line	★						
Faulty nozzle				★	★	★	★
Sticking pump plunger		★		★			
Binding pump control rod				★			
Weak or broken governor springs				★			
Fuel delivery valve not seating properly				★			
Weak or broken transfer pump plunger spring		★	★				
Improperly set smoke stop		★				★	
Broken spring in by-pass valve	★						

day, prior to starting the engine. The fuel then flows to the renewable element type primary filter (B). The drain cock (C) at the bottom of the primary filter should be opened and a small quantity of fuel drained each day, prior to starting the engine. From the primary filter, the fuel passes through the transfer pump (C—Fig. AC61) to the renewable element type intermediate filter (E). If any signs of water are apparent when draining fuel from the primary filter, a small quantity of fuel must be drained from the intermediate filter.

The fuel then passes through the secondary (or final) sealed-type filter (F) and into the injection pump sump. The secondary (or final) filter must be renewed as a unit.

The fuel transfer pump, pumps fuel through the intermediate and secondary (or final) filters under a normal pressure of 15 psi.

147B. To check the filters and/or the transfer pump, install a suitable pressure gage in series between the secondary (or final) fuel filter outlet and the injection pump fuel inlet. Start the engine and observe the pressure gage reading. The gage should register at least 5 psi. If the gage reading is low, renew the intermediate stage element and recheck pressure reading. If the gage reading is still low, renew the complete secondary (or final) filter and recheck the pressure reading. If the gage reading is still low, the transfer pump is not operating properly and same should be renewed and/or overhauled. Refer to paragraph 150A.

INJECTION NOZZLES

WARNING: Fuel leaves the injection nozzles with sufficient force (2000 psi) to penetrate the skin. When testing, keep your person clear of the nozzle spray.

148. **TESTING AND LOCATING FAULTY NOZZLE.** If the engine does

not run properly and the quick checks in paragraph 147 point to a faulty injection nozzle, locate the faulty nozzle as follows:

If one engine cylinder is misfiring, it is reasonable to suspect a faulty nozzle. Generally, a faulty nozzle can be located by loosening the high pressure line fitting on each nozzle holder in turn, thereby allowing fuel to escape at the union rather than enter the cylinder. As in checking spark plugs in a spark ignition engine, the faulty nozzle is the one which, when its line is loosened, least affects the running of the engine .

148A. Remove the suspected nozzle from the engine as outlined in paragraph 149. If a suitable nozzle tester is available, check the nozzle, as in paragraph 148B, 148C, 148D and 148E. If a nozzle tester is not available, reconnect the fuel line and with the nozzle tip directed where it will do no harm, crank the engine with the starting motor and observe the nozzle spray pattern as shown in Fig. 61A.

If the spray pattern is ragged, as shown in the left hand view, the nozzle valve is not seating properly and same should be reconditioned as outlined in paragraph 149A. If cleaning and/or nozzle and tip renewal does not restore the unit and a nozzle tester is not available for further checking, send the complete nozzle and holder assembly to an official diesel service station for overhaul.

148B. **NOZZLE TESTER.** A complete job of testing and adjusting the nozzle requires the use of a special tester such as the American Bosch Nozzle Tester TSE 7722D which is available through any of the Bosch authorized service agencies, etc. The nozzle should

be tested for leakage, spray pattern and opening pressure. Operate the tester lever until oil flows and attach the nozzle and holder assembly.

Note: Only clean, approved testing oil should be used in the tester tank.

Close the tester valve and apply a few quick strokes to the lever. If undue pressure is required to operate the lever, the nozzle valve is plugged and same should be serviced as in paragraph 149A.

148C. LEAKAGE. The nozzle valve should not leak at a pressure less than 1700 psi. To check for leakage, actuate the tester handle slowly and as the gage needle approaches 1700 psi, observe the nozzle tip for drops of fuel. If drops of fuel collect at pressures less than 1700 psi, the nozzle valve is not seating properly and same should be serviced as in paragraph 149A.

148D. SPRAY PATTERN. Operate the tester handle at approximately 100 strokes per minute and observe the spray pattern as shown in Fig. AC61A. If the nozzle has a ragged spray pattern as shown in the left view, the nozzle valve should be serviced as in paragraph 149A.

148E. OPENING PRESSURE. While operating the tester handle, observe the gage pressure at which the spray occurs. The gage pressure should be 2000 psi. If the pressure is not as specified, remove the nozzle protecting cap, exposing the pressure adjusting screw and locknut. Loosen the locknut and turn the adjusting screw as shown in Fig. AC61B either way as required to obtain an opening pressure of 2000 psi. Note: If a new pressure spring has been installed in the nozzle holder, adjust the opening pressure to 2020 psi. Tighten the locknut and install the protecting cap when adjustment is complete.

Fig. AC61 — Fuel filters and model PSB Bosch injection pump installation.
C. Transfer pump   F. Secondary (or final)
E. Intermediate filter   stage filter

Fig. AC61A—Typical spray patterns of a throttling type pintle nozzle. Left: Poor Spray pattern. Right: ideal spray pattern.

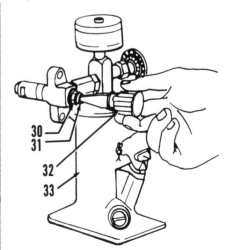

Fig. AC61B — Adjusting nozzle opening pressure, using a nozzle tester.
30. Nut   32. Screw driver
31. Adjusting screw   33. Nozzle tester

149. REMOVE AND REINSTALL. Before loosening any lines, wash the nozzle holder and connections with clean diesel fuel or kerosene. After disconnecting the high pressure and leak-off lines, cover open ends of connections with tape or composition caps to prevent the entrance of dirt or other foreign material. Remove the nozzle holder stud nuts and carefully withdraw the nozzle from cylinder head, being careful not to strike the tip end of the nozzle against any hard surface.

Thoroughly clean the nozzle recess in the cylinder head before reinserting the nozzle and holder assembly. It is important that the seating surfaces of recess be free of even the smallest particle of carbon which could cause the unit to be cocked and result in blowby of hot gases. No hard or sharp tools should be used for cleaning. A piece of wood dowel or brass stock properly shaped is very effective. Do not reuse the copper ring gasket (1—Fig. AC62), always install a new one. Tighten the nozzle holder stud nuts to a torque of 14-16 Ft.-Lbs.

149A. MINOR OVERHAUL OF NOZZLE VALVE AND BODY. Hard or sharp tools, emery cloth, crocus cloth, grinding compounds or abrasives of any kind should NEVER be used in the cleaning of nozzles. A nozzle cleaning and maintenance kit is available through any American Bosch Service Agency under the number of TSE 7779.

Wipe all dirt and loose carbon from the nozzle and holder assembly with a clean, lint free cloth. Carefully clamp nozzle holder assembly in a soft jawed vise and remove the nozzle holder nut and spray nozzle. Reinstall the holder nut to protect the lapped end of the holder body. Normally, the nozzle valve (V—Fig. AC62A) can be easily withdrawn from the nozzle body. If the valve cannot be easily withdrawn, soak the assembly in fuel oil, acetone, carbon tetrachloride or similar carbon solvent to facilitate removal. Be careful not to permit the valve or body to come in contact with any hard surface.

Clean the nozzle valve with mutton tallow used on a soft, lint free cloth or pad. The valve may be held by its stem in a revolving chuck during this cleaning operation. A piece of soft wood well soaked in oil will be helpful in removing carbon deposits from the valve.

The inside of the nozzle body (tip) can be cleaned by forming a piece of soft wood to a point which will correspond to the angle of the nozzle valve seat. The wood should be well soaked in oil. The orifice of the tip can be cleaned with a wood splinter. The outer surfaces of the nozzle body should be cleaned with a brass wire brush and a soft, lint free cloth soaked in a suitable carbon solvent.

Thoroughly wash the nozzle valve and body in clean diesel fuel and clean the pintle and its seat as follows: Hold the valve at the stem end only and using light oil as a lubricant, rotate the valve back and forth in the body. Some time may be required in removing the particles of dirt from the pintle valve; however, abrasive materials should never be used in the cleaning process.

Test the fit of the nozzle valve in the nozzle body as follows: Hold the body at a 45 degree angle and start the valve in the body. The valve should slide slowly into the body under its own weight. Note: Dirt particles, too small to be seen by the naked eye, will restrict the valve action. If the valve sticks, and it is known to be clean, free-up the valve by working the valve in the body with mutton tallow.

Before reassembling, thoroughly rinse all parts in clean diesel fuel and make certain that all carbon is removed from the nozzle holder nut. Install nozzle body and holder nut, making certain that the valve stem is

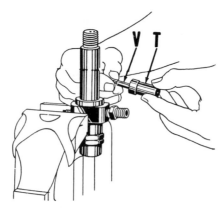

Fig. AC62A — Removing injection nozzle valve (V) from tip (T). If the valve is difficult to remove, soak the assembly in a suitable carbon solvent.

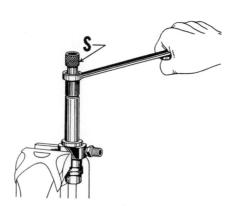

Fig. AC62B—Using Bosch tool (S) to center the nozzle tip while tightening the cap nut.

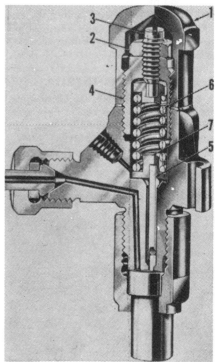

Fig. AC62C — Injection nozzle sectional view.

1. Cap nut	cap nut
2. Jam nut	5. Spindle
3. Adjusting screw	6. Spring
4. Spring retaining	7. Lower spring seat

Fig. AC62 — Sectional view showing the injection nozzle installation. Whenever the nozzle has been removed, always renew the copper gasket (1).

located in the hole of the holder body. It is essential that the nozzle be perfectly centered in the holder nut. A centering sleeve is supplied in American Bosch kit TSE 7779 for this purpose. Slide the sleeve over the nozzle with the tapered end centering in the holder nut. Tighten the holder nut, making certain that the sleeve is free while tightening. Refer to Fig. AC62B.

Test the nozzle for spray pattern and leakage as in paragraph 148C and 148D. If the nozzle does not leak under 1700 psi, and if the spray pattern is symmetrical as shown in right hand view of Fig. AC61A, the nozzle is ready for use. If the nozzle will not pass the leakage and spray pattern tests, renew the nozzle valve and seat, which are available only in a matched set; or, send the nozzle and holder assembly to an official diesel service station for a complete overhaul which includes reseating the nozzle valve pintle and seat.

149B. **OVERHAUL OF NOZZLE HOLDER.** (Refer to Fig. AC62C.) Remove cap nut (1) and gasket. Loosen jam nut (2) and adjusting screw (3). Remove the spring retaining nut (4) and withdraw the spindle (5) and spring (6). Thoroughly wash all parts in clean diesel fuel and examine the end of the spindle which contacts the nozzle valve stem for any irregularities. If the contact surface is pitted or rough, renew the spindle. Examine spring seat (7) for tightness to spindle and for cracks or worn spots. Renew the spring seat and spindle unit if the condition of either is questionable. Renew any other questionable parts.

Reassemble the nozzle holder and leave the adjusting screw locknut loose until after the nozzle opening pressure has been adjusted as outlined in paragraph 148E.

INJECTION PUMP

150. **TIMING TO ENGINE.** The injection pump should be timed so that injection occurs at 21 degrees before top center.

To check and retime the pump to the engine after the pump is installed as outlined in paragraph 152A, proceed as follows:

Crank the engine until No. 1 piston is coming up on compression stroke and the engine flywheel mark "FPI" is in the exact center of the timing hole. Remove plugs (P—Fig. AC63) so that the pump timing marks can be seen. If the pump timing is correct, the line mark (50—Fig. AC63A) on the drive gear hub will be in register with the pointer (51) extending from the front face of the pump. If the timing marks are not in register, remove the cover from front of the

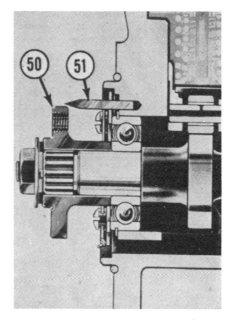

Fig. AC63A—Bosch model PSB injection pump timing mark (50) is a line on edge of pump drive hub. The pointer is shown at (51).

engine timing gear case and loosen the three cap screws retaining the pump drive gear to the pump hub. Using a socket wrench, turn the pump hub until the timing marks are exactly in register and tighten the drive gear retaining cap screws.

150A. **TRANSFER PUMP.** The PSB injection pumps are equipped with a positive-displacement, gear-type transfer pump which is gear driven from the injection pump camshaft. Refer to Fig. AC63B. To check the operation of the transfer pump, refer to paragraph 147B.

If the pump is not operating properly, the complete pump can be renewed as a unit; or, the transfer pump can be disassembled and cleaned and checked for improved performance. Quite often, a thorough cleaning job will restore the pump to its original operating efficiency.

151. **HYDRAULIC HEAD.** The hydraulic head assembly (Fig. AC64) can be renewed without the use of special testing equipment. The head assembly contains all of the precision components which are essential to accurate pumping, distributing, metering and delivery of the fuel. To renew the hydraulic head assembly, first wash the complete injection pump and injection lines with clean fuel oil. Remove the injection lines and disconnect the inlet and outlet lines from

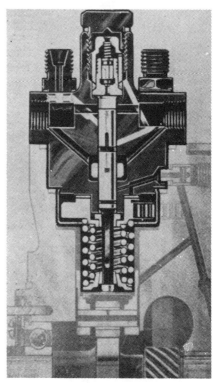

Fig. AC64—Sectional view of Bosch model PSB injection pump hydraulic head. The complete head assembly can be renewed as a unit.

Fig. AC63—To view the Bosch model PSB injection pump timing marks, it is necessary to remove plugs (P).

Fig. AC63B—Cut-away view of model SGB type fuel transfer pump which is used on model PSB injection pumps.

hydraulic head. Remove the timing window cover (21—Fig. AC64A) and crank engine until the line mark on the apex of one of the teeth on the pump plunger drive gear is in register with the "O" mark or arrow stamped on the lower face of the timing window hole as shown in Fig. AC64B. Remove the two screws and carefully withdraw the control assembly, being careful not to lose plunger sleeve pin (24—Fig. AC64A). Unscrew and remove lube oil filter (91). Remove the hydraulic head retaining stud nuts and carefully withdraw the hydraulic head assembly from the pump housing. Do not use force when attempting to withdraw the hydraulic head. If difficulty is encountered, check to make certain that the plunger drive gear is properly positioned as shown in Fig. AC64B.

When installing a new hydraulic head assembly, make certain that the line marked plunger drive gear tooth is in register with the "O" mark or arrow in the timing window as shown

in Fig. AC64B and that open tooth on quill shaft gear is in register with punch mark in pump housing as shown at (A) in Fig. AC65. When installing the control sleeve assembly, the plunger sleeve pin (24—Fig. AC64A) must be lined up with the slot in the control block. The remainder of the reassembly procedure is evident .

152. **REMOVE AND REINSTALL INJECTION PUMP.** Before attempting to remove the injection pump, thoroughly wash the pump and connections with clean diesel fuel. Disconnect the injection lines from injection pump and the inlet and outlet lines from the transfer pump. Disconnect the remaining lines and control rods. Cover all fuel line connections with tape or composition caps to eliminate the entrance of dirt. Crank engine until No. 1 piston is coming up on compression stroke and the flywheel mark "FPI" is in the exact center of the timing hole. Remove

the pump mounting cap screws and withdraw the pump.

152A. To install the injection pump, first make certain that the flywheel mark "FPI" is in the exact center of the timing hole when No. 1 piston is coming up on compression stroke. Loosen the three pump drive gear retaining cap screws. Remove the timing window cover (21—Fig. AC64A) from side of pump housing. Turn the pump hub until the line mark on the apex of one of the pump plunger drive gear teeth is approximately in the center of the timing window. Then, continue turning the hub until the line mark (50—Fig. AC63A) on the drive gear hub is in register with the pointer (51) extending from the front face of the pump. Mount the injection pump on the engine. The pump drive gear should be meshed with the camshaft gear so that the three drive gear retaining cap screws are approximately in the center of the elongated drive gear holes when the

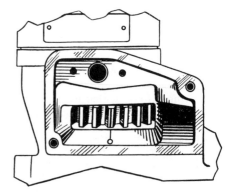

Fig. AC64B—Side view of Bosch model PSB injection pump with timing window cover removed. Notice that the line on one of the gear teeth is in register with the "O" mark or arrow on the housing.

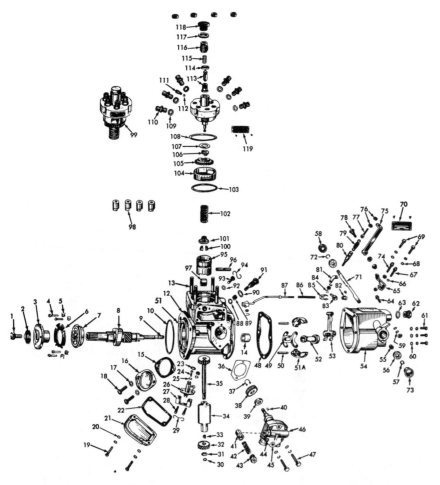

Fig. AC64A — Partially exploded view of Bosch model PSB injection pump.

21. Timing window cover	26. Control unit assembly	88. "O" ring	114. Gasket
24. Sleeve pin	37. Snap ring	89. Filter screen	115. Delivery valve spring
25. Snap ring	38. Gear	90. Gasket	117. Gasket
	39. Seal	91. Filter screw	
	46. Transfer pump	99. Hydraulic head assembly	

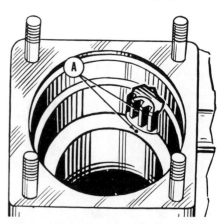

Fig. AC65—When installing a hydraulic head on a model PSB injection pump, make certain that open tooth on quill shaft gear is in register with punch mark in housing as shown at (A).

aforementioned timing marks are in register.

After the pump is installed, make the final timing setting as outlined in paragraph 150.

153. **GOVERNOR.** Model PSB injection pumps are equipped with a mechanical flyweight type governor. For the purposes of this manual, the governor will be considered as an integral part of the injection pump.

153A. ADJUSTMENT. Recommended governed speeds are as follows:

Maximum no load
engine speed ....1975–2000 rpm

Maximum full load
engine speed .... 1625 rpm

Power take-off speed at
1975 engine rpm..... 668 rpm

Low idle engine speed.. 600–650 rpm

To adjust the governor, first start engine and run until engine is at normal operating temperature.

153B. The low idle speed of the engine is controlled by the position of the throttle lever and is adjusted by placing the throttle in the idle position and adjusting the idle stop screw (74—Fig. AC64A) to give 600-650 rpm at low idle.

The shut-off stop screw (67) should be set so that no force can be applied to the control arm in the hydraulic head control sleeve when it is in the shut off position. This screw can be set by using spacers as required. Spacers are available in 0.062, 0.125 and 0.188 thicknesses.

The high idle engine speed is adjusted by nuts (76—Fig. AC66) which vary the length of the externally located governor spring.

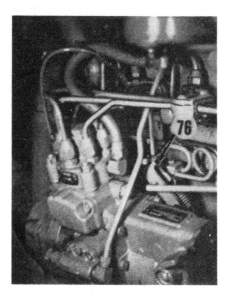

Fig. AC66 — The engine high idle speed is adjusted with nuts (76).

ENERGY CELLS

154. **R&R AND CLEAN.** The necessity for cleaning the energy cells is usually indicated by excessive exhaust smoking, or when fuel economy drops. To remove Nos. 2, 3, 4 or 5 energy cells, remove the air intake pipe and intake manifold. Remove the energy cell clamp and tap the energy cell cap with a hammer to break loose any carbon deposits. Using a pair of pliers remove the energy cell cap. A ¼-inch tapped hole is also provided in the cap to facilitate removal.

The outer end of the energy cell body is tapped with a ⅞-14 thread to permit the use of a screw type puller when removing the cell body. The cell body can also be removed by first removing the respective nozzle, and using a brass drift inserted through the nozzle hole, bump the cell out of the cylinder head.

The removed parts can be cleaned in an approved carbon solvent. After parts are cleaned, visually inspect them for cracks and other damage. Renew any damaged parts. Inspect the seating surfaces between the cell body

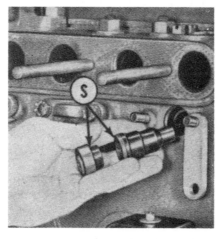

Fig. AC67—Installing the energy cell. If the surfaces (S) are rough or pitted, they can be reconditioned by lapping.

and the cell cap for being rough and pitted. The surfaces (S—Fig. AC67) can be reconditioned by lapping with valve grinding compound. Make certain that the energy cell seating surface in cylinder head is clean and free from carbon deposits.

When installing the energy cell, tighten the clamp nuts enough to insure an air tight seal.

PREHEATER

155. An exploded view of the heater box is shown in Fig. AC68. Normal service includes renewing gaskets and making certain that electrical connections are tight.

DIESEL COOLING SYSTEM

RADIATOR

156. To remove radiator, first drain cooling system. Remove hood and loosen top and bottom radiator hoses. Remove cap screws holding radiator and shell to frame and remove radiator and shell as a unit. Further disassembly of radiator and shell is self evident upon examination. Reinstall by reversing the above procedure.

THERMOSTAT

157. The thermostat can be removed after removing the thermostat housing from front of the water outlet casting which is located on top of cylinder head. The thermostat is preset to open at 170 degrees.

WATER PUMP

158. **RESEAL AND OVERHAUL.** The pump cannot be overhauled without removing the pump from the trac-

Fig. AC68 — Exploded view of engine pre-heater unit.

13. Heater element preheat-chamber
14. Element spacer
15. Screen

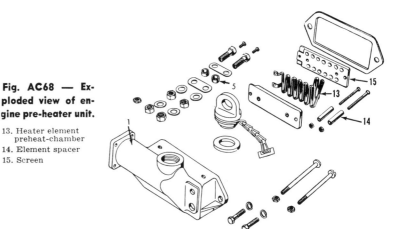

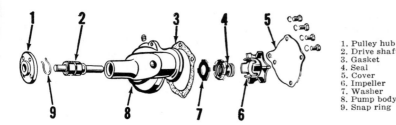

1. Pulley hub
2. Drive shaft
3. Gasket
4. Seal
5. Cover
6. Impeller
7. Washer
8. Pump body
9. Snap ring

Fig. AC69—Exploded view of water pump. The pump can be removed after removing the radiator.

DIESEL ELECTRICAL SYSTEM

GENERATOR, REGULATOR AND STARTING MOTOR

159. Delco-Remy generator No. 1100994, regulator No. 1118791 and starting motor No. 1108998 are used. Test specifications are as follows:

1100994 Generator

Brush spring tension	16 oz.
Field draw, volts	12.0
Field draw, amperes	2.0-2.14
Hot output, amperes	9-11
Hot output, volts	13.8-14.2
Hot output, rpm	2400

1118791 Regulator

Cut-out relay

air gap	0.020
point gap	0.020
closing voltage—range	11.8-14.0
—adjust to	12.8

Voltage regulator

air gap	0.075
voltage setting—range	14.0-15.0
—adjust to	14.4
Ground polarity	P

1108998 Starter

Volts	12
Brush spring tension	36-40 oz.
No load test, volts	11.3
No load test, amperes	65
No load test, rpm	5500
Lock test, volts	4.0
Lock test, amperes	675
Lock test, torque, ft.-lbs.	30

tor. To remove the pump, first remove the radiator. Remove fan blades and pump pulley and unbolt pump from cylinder head. Using a press or suitable puller, remove pulley hub. Remove cover from back of pump and using a pair of long nose pliers, remove the pump shaft retaining snap ring. Press the drive shaft and bearing unit out of impeller and pump body.

Renew bearing and shaft unit (2—Fig. AC69) if shaft is bent or if bearing is a loose fit in pump body. Inspect seal contacting surface in pump body. If the surface is pitted or rough, either recondition the surface or re-new the pump body. Always renew a questionable seal.

When reassembling, press the shaft and bearing unit into the pump body and install the retaining snap ring. Assemble seal into impeller and start the impeller on the shaft. Place pump in a vise and press the impeller on the drive shaft until end of shaft is flush with rear surface of the impeller. Press pulley hub on front of drive shaft and install cover.

Note: A newly overhauled pump might leak for the first few hours of operation. This condition should correct itself after the parts have had an opportunity to wear in.

CLUTCHES

CLUTCH ON ENGINE

Models B-C-CA

Rockford clutch model 9TT with R2-3271 cover was used in model B tractors up to serial number B7134; model 9RM with R4409 cover in tractors after serial number B7134 to engine number 28222; 8½RM with R4665 cover in engine number 28222 and up, and in models C and CA tractors. Some early model B tractors were equipped with Auburn Atwood clutches. Front end of clutch shaft is piloted into an Oilite bushing in the flywheel. Pilot bearing lubricating wick is renewable; refer to Crankshaft, paragraph 97, for procedure.

160. **ADJUSTMENT.** Remove inspection plate (15—Fig. AC72) from bottom of clutch housing or torque tube. Turn release lever adjusting screws (17) to obtain ¼ inch clearance between ends of release levers and re-lease bearing (16). Clearance between release bearing and each lever must be the same within .010 inch.

Three threads of models B, C and CA clutch rod (13) should protrude through clutch fork pin (3). Some early B model tractors were equipped with Atwood clutches, and on these tractors the clutch rod protrudes only one thread through the fork pin. If a Rockford clutch is installed in these tractors, be sure to turn the rod until three threads protrude through the fork pin.

161. **R&R CLUTCH UNIT.** Place supports under rear of engine and center of torque tube. Loosen head light mounting bolts. Remove hood and disconnect fuel line and governor control rod. Disconnect rear end of radius rod and drag link. Remove starting motor and air cleaner. Remove bolts retaining engine to torque tube and separate tractor halves. Clutch can be removed by removing cap screws retaining clutch cover to flywheel.

Reassembly is reverse of disassembly. Install driven plate with oil deflector toward flywheel. Align driven plate with a clutch pilot tool or clutch shaft.

162. **RELEASE BEARING.** After engine has been detached from torque tube, disconnect clutch rod (13—Fig. AC72) from clutch pedal and from clutch fork pin (3). Remove shifter shaft retainer (5) from between shifter fork (4) and torque tube. Remove shifter fork shaft (1) and slide clutch shifter (2), release bearing (16), and shifter fork off clutch shifter tube (14). Release bearing can be pressed off shifter.

163. **CLUTCH SHAFT.** To remove clutch shaft (8) at this time, it is necessary to first detach torque tube from transmission, as outlined in paragraph 185A. Clutch shaft can be withdrawn from rear end of torque tube.

1. Clutch shifter fork shaft
2. Clutch shifter
3. Clutch fork pin
4. Clutch shifter fork
5. Shifter shaft retainer
6. Clutch cover
7. Pressure plate
8. Clutch shaft
9. Torque tube
10. Clutch pedal
11. Universal joint
12. Universal joint retaining pin
13. Clutch rod

14. Clutch shifter tube
15. Inspection plate
16. Release bearing
17. Release lever adjusting screw
18. Oil deflector
19. Driven plate
20. Clutch cover retaining cap screw
21. Clutch shaft pilot bushing

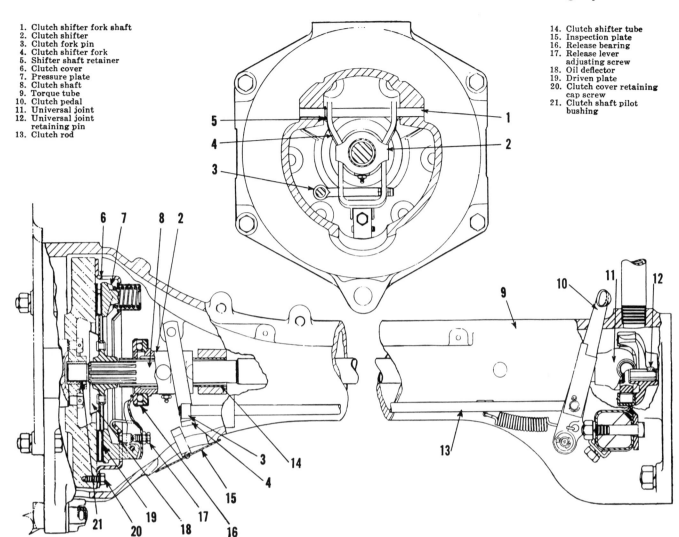

Fig. AC72 — Early production models B, and C clutch shifting mechanism, clutch shaft, and torque tube assembly. On later B and CA models, the clutch shaft is connected to the transmission main shaft by means of a splined coupling (sleeve) and cotter key instead of a universal joint.

Fig. AC73—Models B, C, and CA showing clutch installation to engine flywheel when torque tube is detached from engine.

6. Clutch cover
7. Pressure plate
17. Release lever adjusting screw
22. Pressure spring
23. Spring cup
24. Release lever

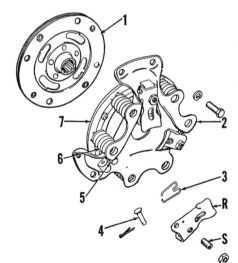

Fig. AC74—Model G clutch is a single plate Rockford 6½ RM unit.

R. Release lever
S. Adjusting screw
1. Driven disc
2. Back plate (cover)
3. Release lever spring
4. Pivot pin
5. Adjusting screw contacting surface
6. Pressure spring
7. Pressure plate

Model G

This tractor is equipped with a Rockford 6½ RM clutch fitted with Rockford cover assembly 165256.

164. **ADJUSTMENT.** With clutch fully engaged, the clutch pedal should have one inch of free travel. This amount of free travel produces the desired 1/16 inch clearance between release bearing face and face of the release levers. This adjustment is made by turning the nut at end of operating rod. Pedal should be approximately 4 inches below seat attaching pad.

165. **R&R CLUTCH.** To remove the unit, it is first necessary to remove the engine as outlined under R & R ENGINE WITH CLUTCH section. The release fork and bearing can be renewed at this time. The unit can now be removed from the flywheel after removing the 6 cap screws. If a tapered block is placed between outer end of each release lever and backing plate before loosening the 6 attaching

screws, it will relieve the spring pressure and facilitate removal.

The long hub of the lined plate (1—Fig. AC74) should be installed away from the flywheel. Using 3 pieces of 5/16 inch key stock in lieu of the lined plate, install the unit to the flywheel and check the release lever height. Adjust release levers, by means of the screws (S) in the levers, until a measurement of 1 13/16± 0.015 is obtained from friction face of flywheel to release bearing contacting surface of each lever.

If the clutch shaft or clutch shaft bearing need overhauling, it will be necessary to remove the engine and clutch housing from the transmission as outlined in paragraph 65.

Models RC-WC-WF

Rockford clutch model 9RM with R4409 cover, is used in model RC tractors. WC tractors prior to number WC74330 and WF tractors prior to number 1337 used Rockford clutch model 10RR with R1716-2 cover. Subsequent WC and WF tractors use clutch model 10RM with R4259 cover.

Clutch shaft pilot bearing is lubricated by a wick in the crankshaft rear main bearing journal. Refer to Crankshaft, paragraph 97, for wick removal procedure.

166. ADJUSTMENT. Method of adjustment is the same as for models B, C, and CA, as outlined in paragraph 160.

167. R&R CLUTCH UNIT. Remove engine from tractor as in paragraph 63 or 64. With engine removed, clutch can be removed by removing cap screws (3—Fig. AC75) retaining clutch cover to flywheel. Reinstall driven plate with oil deflector of same toward the transmission.

SEGMENT LINING TYPE. Each side of the driven plate has two 1/8 inch thick segments alternated by two 5/32 inch thick segments. The 1/8 inch thick segments of one side are riveted to the 5/32 inch thick segments on the other side. Install rivets with the heads in the 1/8 inch thick segments.

FULL CIRCLE LINING TYPE. Rivet molded lining to flywheel side of driven plate and woven lining to the pressure plate side.

Install driven plate with oil deflector toward transmission.

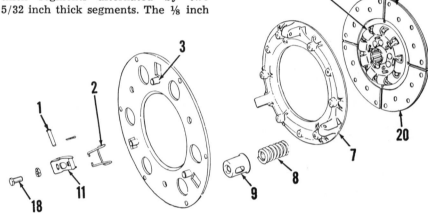

Fig. AC76—Some WD and WD45 tractors are equipped with a Rockford 10 RM unit which has a segment type lined disc. Rivets are installed with their heads in the 1/8 inch segments. Some RC, WC, and WF models are similar.

1. Lever pivot pin	4. Driven disc	8. Pressure spring	11. Release lever
2. Lever spring	5. 5/32 inch lining	9. Spring cup	18. Adjusting screw
3. Screw seat	7. Pressure plate		20. 1/8 inch lining

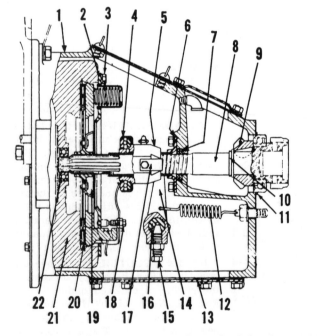
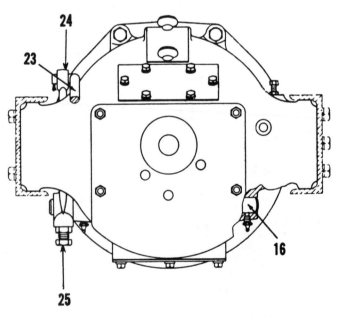

Fig. AC75—Models RC, WC, and WF clutch, clutch housing assembly. Models WD and WD45 are similar with the exception of clutch shaft (8); refer to Fig. AC84 for difference.

1. Clutch housing	6. Shifter bracket	11. Snap ring	16. Shifter fork shaft	21. Flywheel
2. Clutch cover plate	7. Clutch shaft oil seal	12. Shifter spring	17. Shifter clip	22. Shaft pilot bearing
3. Cap screw	8. Clutch shaft	13. Inspection plate	18. Adjusting screw	23. Control rod
4. Release bearing	9. Belt pulley drive gear	14. Shifter fork	19. Oil deflector	24. Shifter fork shaft arm
5. Clutch shifter	10. Snap ring	15. Shifter fork set screw	20. Driven plate	25. Set screw

168. RELEASE BEARING. After engine has been detached from clutch housing, disengage shifter (5—Fig. AC75) from shifter fork (14) and remove shifter and release bearing (4) from shifter bracket (6). Release bearing can be pressed off shifter.

169. CLUTCH SHAFT. For repair information, refer to paragraph 226.

170. SHAFT SEAL. With release bearing removed, remove set screw (15) from shifter fork and disconnect spring (12). Turn shifter fork down, remove cap screws from shifter bracket (6) and remove bracket and cork seals (7).

171. SHIFTER FORK. Can be removed from below without removing engine by removing shifter fork shaft arm (24), extracting shaft (16) from housing, and withdrawing fork from below.

Models WD-WD45 Engine Clutch

Models WD and WD45 tractors are equipped with either a 10 inch single plate spring loaded model 10RM Rockford clutch or an A10 Auburn clutch. Clutch cover assembly number is: Rockford, R4259; Auburn 100057-1.

Flywheels used on models equipped with a Rockford clutch are not interchangeable with the same tractor model which is equipped with an Auburn clutch.

Effective with the WD tractor equipped with engine W310920 and all WD45 tractors, the clutch shaft has no pilot bearing.

For repair information on clutch shaft, shaft bearings, or shaft rear oil seal, refer to TORQUE TUBE.

174. ADJUSTMENT. Clutch pedal should have ½ to one inch of free travel when measured between pedal and top of tractor frame rail. This amount of free pedal travel produces the desired clearance of ¼ inch between release bearing face and face of release levers. Adjustment is made by turning the threaded pivot (12—Fig. AC77).

175. R&R & RELINE. To remove the clutch, first remove the engine assembly as outlined under R & R ENGINE WITH CLUTCH. Insert a tapered block between outer end of each release lever and backing plate to relieve spring pressure and remove the six clutch retaining cap screws.

Each side of the Rockford clutch driven plate has two ⅛ inch thick segments (20—Fig. AC76) alternated by two 5/32 inch thick segments (5). The ⅛ inch segments of one side are riveted to the 5/32 inch segments of the other side. Install all rivets with the

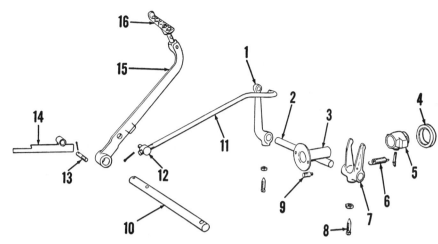

Fig. AC77—Models WD and WD45 engine clutch release bearing, fork and linkage—exploded view. Clutch pedal free travel adjustment is made at (12). Models RC, WC, and WF are similar.

1. Fork lever	5. Bearing sleeve	9. Spring anchor	13. Latch stud
2. Fork shaft	6. Fork return spring	10. Clutch pedal shaft	14. Pedal latch
3. Sleeve bracket	7. Shifter fork	11. Clutch rod	15. Clutch pedal
4. Release bearing	8. Set screw	12. Adjusting pin	16. Pedal pad

Fig. AC78 — Model WD transmission clutch installation. Clutch assembly can be removed after detaching transmission from torque tube. To adjust clutch, add or remove shims (1). Note hydraulic pump hold valve adjusting screw at (C). Model WD45 is similar.

rivet heads in the ⅛ inch thick segments.

176. FRONT OIL SEAL. The clutch housing contains the engine clutch shifter fork (7—Fig. AC77), clutch release bearing (4) and clutch shaft front oil seal (9—Fig. AC84). The shifter fork can be removed from below without detaching engine from clutch housing. However, renewal of the clutch release bearing and/or sleeve will require detaching engine from clutch housing.

Clutch shaft front oil seal (9) of treated leather can be renewed after detaching engine from clutch housing and removing release bearing guide sleeve. Install new oil seal with lip facing rearward (transmission).

176A. CLUTCH SHAFT. For repair information, refer to Torque Tube, paragraph 192.

177. R&R CLUTCH HOUSING. Proceed as follows: Remove engine as outlined under R & R ENGINE WITH CLUTCH. Block up and support tractor, and remove the left main frame rail. Remove fuel tank, air cleaner and starting motor. From the right main

frame rail, remove belt pulley unit and two clutch housing retaining bolts. Disconnect torque tube from clutch housing and remove clutch housing.

CA FINAL DRIVE CLUTCH

178. An auxiliary clutch, mounted in series with the right final drive unit bull pinion shaft, is used on this model when it is sold with the A-C continuous power PTO. For repair information on this unit, refer to paragraphs 284 through 285A.

WD-WD45 TRANSMISSION CLUTCH

The transmission clutch is an 8-inch Rockford over-center, double disc type, as shown in Fig. AC78. The clutch plates are faced with metallic type friction material.

The transmission clutch is installed only in those tractors equipped with the continuous power PTO.

179. ADJUSTMENT. To compensate for disc wear, proceed as follows: Remove battery, battery carrier and the clutch housing (torque tube) top cover. With the clutch disengaged and front plate (8—Fig. AC78), center plate (7) and driven discs (9) moved forward, check the clearance between the pressure plate (11) surface and rear face of driven disc. Desired clearance of 0.030 is obtained by adding or removing shims (1) from the three adjusting points.

180. **OVERHAUL.** To remove clutch, proceed as follows: Drain torque tube housing. Disconnect clutch pedal, brake rods and tail light wire. Remove battery, battery carrier, transmission clutch housing (torque tube) top cover and power take-off shaft. Remove steering post-to-transmission housing retaining cap screws. Disconnect both hydraulic lift rams. Engage transmission clutch to facilitate its removal through top opening in torque tube. Block up and support tractor under rear portion of main frame. Support torque tube and transmission housing separately. Remove torque tube-to-transmission housing retaining bolts and roll transmission assembly away

from tractor. Working through opening in top of torque tube, remove four nuts attaching transmission clutch front plate (8—Fig. AC79) to flanged power take-off drive gear (5—Fig. AC84). Lift transmission clutch out through top.

Reassemble the clutch unit so that the long hub of forward disc faces the engine and the long hub of the rear disc faces the transmission.

The Belleville type spring washer (2—Fig. AC79) should be installed with its small diameter facing forward.

The bi-metallic friction facings are integral with the driven plates, and are sold only as a plate assembly. They should be renewed when worn to the point where the grooves are obliterated.

181. **CLUTCH SHAFT.** For repair information on transmission clutch shaft (7—Fig. AC83 or AC104) refer to paragraph 233.

TORQUE TUBE
Models B-C-CA

185. To remove torque tube (clutch housing), Fig. AC80, from tractor, first place supports under torque tube and under the engine. Remove hood, starting motor, air cleaner, and fuel tank strap. Disconnect steering drag link, radius rod on axle models, and wiring harness at cutout relay, ignition unit, and head lights. Remove nuts from four bolts which retain torque tube to engine.

185A. On B and C, unbolt front end of each wheel guard from crosswise support and disconnect brake linkage. On CA, remove wheel guards and wheel guard supports and disconnect brake linkage.

On all models, disconnect tail light wire and remove nuts from studs which retain torque tube to transmission. Move transmission back far enough to permit removing the cotter pin or straight pin which retains universal joint to the transmission shaft. After extracting the pin, the torque tube with steering gear panel box and other parts can be moved away from the tractor.

If the torque tube is to be renewed, remove the steering gear, battery, metal shroud, and tool box. Remove clutch shaft, clutch shifter fork, brake pedal pivot shaft, and other parts.

Models RC-WC-WF (Without Traction Release Coupling)

186. To remove torque tube, support rear of tractor frame and detach rear axle housing from end of each frame side rail. (On WF model, remove also the platform and steering gear.) Remove steering support brace and roll

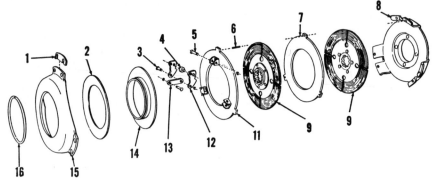

Fig. AC79—Model WD transmission clutch assembly is an overcenter type. Disc wear is compensated for by removal of shims (1). Model WD45 is similar.

1. Shims	5. Pivot pin	9. Driven disc	14. Belville retainer
2. Belville washer	6. Rear plate spring	11. Pressure plate	15. Rear plate
3. Link pin	7. Center plate	12. Release lever	16. Snap ring
4. Release lever roller	8. Front plate	13. Lever link	

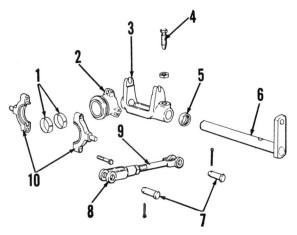

Fig. AC79A—Model WD transmission clutch shifter & bushing, fork and linkage, which are located in rear portion of torque tube. Model WD45 is similar.

1. Shifter bushings	6. Shifter shaft
2. Shifter & bushings	7. Yoke pin
3. Fork	8. Link yoke
4. Set screw	9. Lever link
5. Oil seal	10. Shifter collar

Fig. AC80—Model CA torque tube installation. Models B, and C are similar.

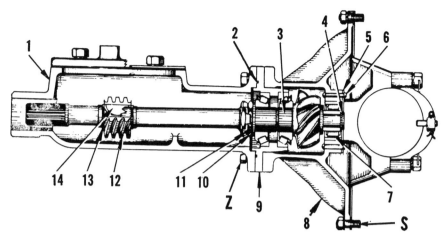

Fig. AC81—Models RC, WC and WF torque tube and drive pinion assembly without traction release coupling. Model RC is not provided with straddle-mounted bearing (7).

2. Cork seal	5. Bearing retainer	8. Differential carrier	12. Power lift worm
3. Drive pinion	6. Retainer rivet	9. Pinion bearing cage	13. Snap ring
4. Rear bearing inner race snap ring	7. Drive pinion rear bearing (not on RC)	10. Bearing adjusting nut	14. Worm key

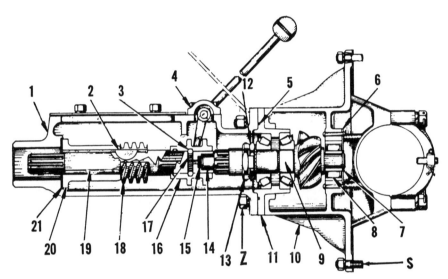

Fig. AC82—Models RC and WC torque tube and drive pinion assembly with traction release coupling. Model RC is not provided with straddle-mounted pinion bearing (8).

1. Torque tube	7. Rear bearing inner race snap ring	11. Pinion bearing cage	16. Shift coupling
2. Worm key	8. Drive pinion rear bearing (not on RC)	12. Adjusting nut	17. Detent ball spring
3. Detent ball	9. Drive pinion	13. Adjusting nut lock	18. Power lift worm
4. Shifter cover	10. Differential carrier	14. Intermediate shaft bearing	19. Intermediate shaft
5. Cork seal		15. Thrust washers	20. Worm spacer
6. Bearing retainer			21. Gear thrust washer

Fig. AC83—Model WD engine power line distribution to the various units. Model WD45 is similar.

A. Engine clutch housing
B. Torque tube
C. Transmission & differential housing
1. Engine clutch
2. Engine clutch shaft
3. Belt pulley drive gear
4. Hydraulic pump drive
5. PTO drive gear
6. Transmission clutch
7. Transmission clutch shaft
8. Bevel pinion shaft
9. Bevel ring gear
10. PTO extension shaft
11. Countershaft
12. PTO intermediate drive gear
13. PTO gear housing
14. PTO gear

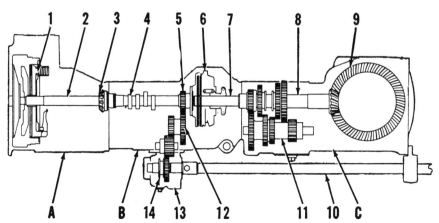

axle and torque tube assembly backward until clear. Remove screws (Z) retaining torque tube (1—Fig. AC81) to differential carrier (8) and slide torque tube off pinion shaft. Main drive bevel pinion shaft and bearing cage assembly can now be withdrawn from the carrier.

187. To disassemble pinion shaft assembly, remove front snap ring (13) and press worm gear off shaft. Remove worm gear key (14), rear snap ring, bearing adjusting nuts (10) and adjusting nut lock (11). Bump pinion shaft (3) rearward through bearing cage (9), removing front bearing cone from front of cage. Rear bearing cone can be pressed off shaft and bearing cups can be driven out of bearing cage. Rear pinion bearing (7) can be removed from WC and WF tractors after removing the differential.

Reassembly is reverse of disassembly. Adjust pinion bearings by tightening adjusting nuts until all bearing play is eliminated, but without binding pinion shaft. Renew cork seal (2) when reinstalling assembly.

Model WC (With Traction Release Coupling Effective WC74330)

188. To remove torque tube, support rear of tractor frame and detach rear axle housing from end of each frame side rail. Roll axle and torque tube assembly backward until clear. Remove coupling shift lever cover (4—Fig. AC82) from torque tube. Remove screws and nuts (Z) retaining torque tube (1) to the differential carrier (10) and slide torque tube off shaft. Intermediate shaft assembly can be removed from pinion shaft (9). After the bronze thrust washer (21) and sleeve spacer (20) are removed from shaft, the power lift worm gear (18) and key can be removed. Needle bearing (14) and steel and bronze thrust washers (15) can be removed from pocket in rear end of intermediate shaft (19). Splined collar (16) can be

pulled from rear end of shaft and detent ball (3) and spring (17) removed. Refer to paragraph 187 for pinion shaft and bearing cage information.

Reassembly is reverse of disassembly. Various thicknesses of bronze and steel thrust washers (15) are available. Select the proper sizes to minimize end play in intermediate shaft. Renew cork seal (5) when reassembling.

Models WD-WD45

190. Transmission clutch (6—Fig. AC83), engine clutch shaft (2), power take-off drive (5) and intermediate gear (12) are contained in the torque tube. However, this tractor model can be equipped in one of three ways: (a) with gear and flange for driving power take-off unit and for connecting transmission clutch as shown; (b) with flange only for connecting transmission clutch; (c) with splined coupling sleeve where neither power take-off unit nor transmission clutch is used.

BASIC PROCEDURE. Although most repair jobs will involve overhaul of the complete unit, there are infrequent instances where the failed or worn part is so located that the repair work can be completed safely without complete disassembly of the torque tube. In effecting such localized repairs, considerable time will be saved by observing the following as a general guide.

Clutch Shaft. Removal of engine clutch shaft (2) requires detaching clutch housing from torque tube and torque tube from transmission.

Transmission Clutch. Removal of clutch (6) requires detaching trans-

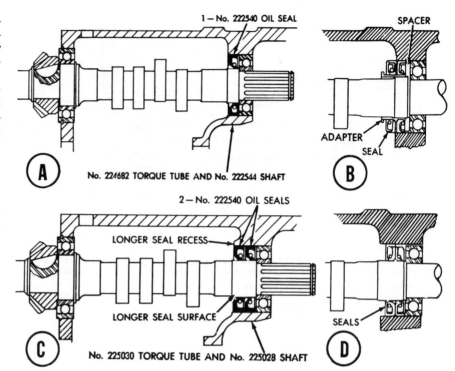

Fig. AC85—Various arrangements used for sealing rear of engine clutch shaft on Model WD. Model WD45 installation is shown in view D.

A—Torque tube and clutch shaft installation on tractors prior to serial 70966. One oil seal was used and installed as shown.

B—The above 225869 seal assembly consisting of an adaptor, spacer, and double seal (lips facing opposite to each other) can and should be installed on tractors prior to serial 70966 which are equipped with only one seal and the old style torque tube and clutch shaft, View A.

C—Tractors within the serial range of 71146 to 72138 and all tractors after 75645 are equipped with the new torque tube and clutch shaft, which is fitted with two 222540 oil seals as shown. Tractors within the serial ranges of 70966 to 71146, and 72138 to 75645, although equipped with the new torque tube and clutch shaft, are factory equipped with only one oil seal. When servicing all models which have the new torque tube and clutch shaft, install two 222540 oil seals with the lips opposite to each other as shown in View D.

D—The two 222540 oil seals should be installed as shown (lips opposite to each other) on all tractors after 70966 or all tractors equipped with the new torque tube and clutch shaft. Sealing of Model WD45 engine clutch shaft is similar.

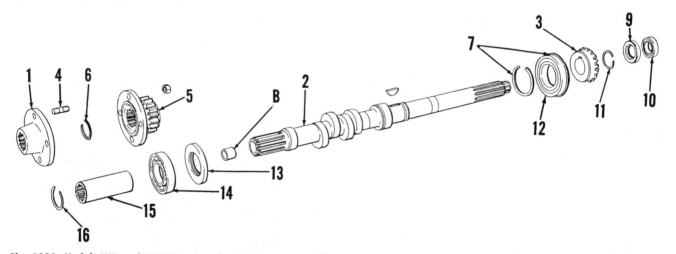

Fig. AC84—Models WD and WD45 engine clutch shaft. Coupling flange (1) is used only when tractor is not equipped with transmission clutch. Coupling sleeve (15) is used on tractors which have neither a transmission clutch nor a power take-off. Refer to Fig. AC85 for various versions or rear oil seals (13). Beginning with WD engine W310920 and all WD45, the clutch shaft has no pilot bushing or bearing (10).

B. Main drive gear shaft pilot bushing	2. Clutch shaft	5. PTO drive gear	10. Engine clutch shaft pilot bearing	12. Bearing	15. Coupling sleeve
1. Coupling flange	3. Belt pulley drive gear	6. Snap ring	11. Snap ring	13. Oil seal	16. Sleeve snap ring
	4. Flange stud	9. Oil seal		14. Clutch shaft rear bearing	

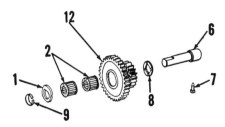

Fig. AC86—Models WD and WD45 power take-off drive intermediate gear and shaft.

1. Thrust washer (rear)
2. Roller bearing
6. Shaft
7. Shaft set screw
8. Thrust washer (front)
9. Spacer
12. Intermediate gear

mission housing from torque tube.

PTO Gears. Removal of power take-off drive gear (5) requires detaching transmission housing from torque tube and removal of transmission clutch.

Removal or overhaul of power take-off intermediate drive gear (12) requires detaching transmission housing from torque tube and detaching torque tube from clutch housing and removal of hydraulic lift pump and transmission clutch.

B.P. Drive Gear. Removal of belt pulley drive gear (3) (located near forward end of clutch shaft) requires removal of engine clutch shaft.

191. **R&R TORQUE TUBE.** First drain torque tube housing. Disconnect clutch pedal rod, brake rods and tail light wire. Remove battery, battery carrier, power take-off shaft and hydraulic pump unit. Disconnect hydraulic lift arms. Remove steering post-to-transmission retaining capscrews. Block up and support tractor under rear portion of frame. (Engage transmission clutch to facilitate re-installation of transmission drive shaft). Support torque tube and transmission with a wheel type jack and remove torque tube-to-clutch housing retaining bolts. Roll transmission and torque tube assembly rearward and away from tractor. Support transmission and torque tube separately and detach torque tube from transmission.

192. **ENGINE CLUTCH SHAFT, SEALS, BEARINGS AND BELT PULLEY DRIVE GEAR.** Removal of engine clutch shaft (2—Fig. AC84) and/or renewal of the clutch shaft rear oil seals or bearings, requires removal of torque tube as outlined in preceding paragraph.

For models equipped with a transmission clutch and power take-off unit proceed as follows: Remove the torque tube. Unbolt and remove the transmission clutch from the gear (5). Remove snap ring and bump gear (5) from shaft. Bump engine clutch shaft forward (towards engine) and out of torque tube.

Clutch shaft single or double rear oil seal (13) can now be renewed as per Fig. AC85. The clutch shaft front oil seal (9) can be renewed at this time after removing the clutch housing lower inspection plate and release bearing guide sleeve. Install front oil seal with lip facing transmission.

Belt pulley drive gear (3) and clutch shaft front bearing (12) can be renewed at this time after removing the snap ring and pressing gear and bearing from shaft.

Running clearance of transmission main shaft bronze pilot bushing (B—Fig. AC84) (located in rear end of clutch shaft) should not exceed .005.

193. **PTO DRIVE GEARS.** Removal of drive gear (5) requires detaching transmission from torque tube and removal of transmission clutch. Remove power take-off gear retaining snap ring (6) and bump gear from shaft.

Removal of power take-off intermediate drive gear (12—Fig. AC 86) requires torque tube removal. Remove intermediate gear shaft retaining set screw (7) (located on outside of torque tube) and bump shaft (6) forward. Remove shaft through hydraulic pump opening in torque tube. Intermediate gear thrust washers (1 & 8) and spacer (9) are removed through opening in lower side of torque tube.

MODEL "B"

TRANSMISSION

Models B-C

The transmission gearset, final drive bevel pinion, bevel ring gear, and differential are carried in the one housing as shown in Fig. AC88. Mesh position of the bevel pinion shaft is fixed and non adjustable. Operations other than R&R control cover require disconnecting transmission from torque tube as outlined in paragraph 185. Operations involving removal of countershaft or idler gear shaft require R&R of differential as per paragraph 241.

All internal operations require R&R of control cover.

201. **COVER, SHIFTER RODS (RAILS) AND FORKS.** The control cover can be removed after draining lubricant by removing cap screws retaining cover to right side of transmission case. The shifter lever can be removed after removing dust cap by removing snap ring (23) from recess in transmission case. A sharp pull up on the lever may be required to dislodge pilot washer (22).

To remove shifter forks cut bottom off pins (3) and drive pins out of holes. Extract rails from cover and forks being careful not to lose detent balls from holes inside forks. When reassembling, install detent springs and balls in shifter forks and slide rails through cover and forks.

202. **MAINSHAFT.** To remove this shaft (13), first disconnect torque tube from transmission as in paragraph 185, and remove power take-off or rear cover plate. Next, remove front bearing retainer (11—Fig. AC88) and

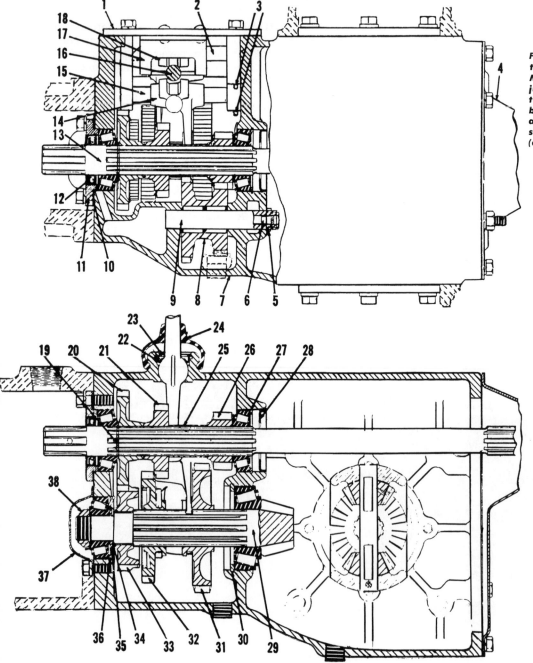

Fig. AC88—Models B, and C transmission assembly. Mainshaft bearings are adjusted with shims (10). Countershaft (bevel pinion shaft) bearings are adjusted with adjusting nut (38). Mesh position of the bevel pinion (countershaft) is fixed and non-adjustable.

2. Second and third shift rail
3. Shift rail lockpins
4. Rear cover plate
6. Idler gear shaft plug
7. Transmission housing
8. Reverse idler gear cluster
9. Reverse idler gear shaft
10. Shims
11. Bearing retainer
12. Oil seal
13. Mainshaft
14. 1st & reverse shift fork
15. 1st & reverse shift rail
16. Shift lever
17. Shift lever guide
18. 2nd & 3rd shift fork
19. Snap ring
20. 3rd speed gear
21. 2nd speed gear
22. Shift lever pilot washer
23. Snap ring
24. Shift lever dust cap
25. Gear (bearing) spacer
26. 1st speed gear
27. Snap ring
28. Mainshaft rear bearing oil cup
29. Countershaft and main drive bevel pinion
30. Snap ring
31. 1st & reverse sliding gear
32. 2nd & 3rd sliding gear
33. 3rd speed constant mesh gear
34. Locating washer
35. Spring washer
36. Snap ring
37. Bearing cap
38. Adjusting nut

bump shaft forward. As shaft and front bearing cone and cup are withdrawn from front of case, gears and rear bearing cone can be removed through side of case. Rear bearing cup is positioned in bore of case by snap ring (27). To remove rear bearing cup, remove oil cup (28) from rear of case and bump bearing cup forward.

Reassembly is reverse of disassembly. Insert mainshaft through front of case and install gears on shaft as follows: Third speed gear (20) with long part of hub toward the rear; second speed gear (21) with long part of hub toward the front; then bearing spacer (gear spacer) (25). Install rear bearing cone in cup and hold in place with first speed gear (26) installed with long part of hub toward the front, then push mainshaft through first speed gear and rear bearing cone. Select shims (10) between front bearing retainer and case to permit shaft to revolve freely with no play. Removing shims reduces play.

203. COUNTERSHAFT (BEVEL PINION). The combination countershaft and bevel pinion (29—Fig. AC88) is supported on two tapered roller bearings and is sold separately from the mating bevel ring gear. Countershaft can be removed after first disconnecting the torque tube from transmission as outlined in paragraph

185, and removing differential as outlined in paragraph 241. Remove bearing cap (37) and bearing adjusting nut (38). Bump countershaft rearward into differential housing and withdraw shaft from housing, gears through side of transmission case and front bearing cone from front of transmission case. Front bearing cup is positioned in bore of the case by snap ring (36) and can be driven out through front of case. Rear bearing cup is positioned in bore of case by snap ring (30) and can be driven out through rear of case.

Reassembly is reverse of disassembly. Install rear bearing cone on countershaft and insert countershaft through rear of case and install gears on shaft as follows: First and reverse sliding gear (31) with groove toward the front; second and third sliding gear (32) with groove toward the rear; third speed constant mesh gear (loose gear) (33) with small diameter teeth toward the rear; locating washer (34) with oil grooves toward gear; then spring washer (35). Install front bearing cone and tighten bearing adjusting nut until shaft revolves freely with no play. Lock bearing adjustment by flattening side of nut into groove in shaft.

204. REVERSE IDLER GEAR SHAFT (9) can be removed after disconnecting torque tube from transmis-

sion and removing differential and transmission mainshaft. Remove cap screw and lock plate (5). Withdraw shaft from the rear and remove gear cluster (8) and straight roller bearings through side of case. A drilled passage in the shaft provides lubrication for the bearings and is sealed at the rear end by plug (6).

Reassembly is reverse of disassembly. Install idler gear cluster with large gear toward the front.

Model CA

205. BASIC PROCEDURES. The transmission shafts, differential unit, final drive bevel pinion, and ring gear are carried in the one transmission and differential housing, Fig. AC92. The mesh position of the bevel pinion is fixed and non adjustable. There are infrequent instances where the failed or worn transmission part is so located that the repair can be completed safely without complete disassembly of the transmission. In effecting such localized repairs, time will be saved by observing the following as a general guide:

Shifter Rails and Forks. Shifter rails and forks, which are located on the transmission cover, are accessible for overhaul after removing the transmission cover.

Main Shaft. Main shaft bearings can be adjusted with shims (2) located between front face of transmission housing and bearing retainer. Adjustment can be made after detaching transmission from torque tube as in paragraph 185. Removal of main shaft requires the additional work of removing the PTO (accessory units) housing from rear of transmission.

Bevel Pinion Shaft. Mesh position of bevel pinion shaft is non-adjustable. Bevel pinion and shaft can be purchased separately from the bevel ring gear. To adjust the bearings or to remove the shaft, it is necessary to detach the torque tube from the transmission, and remove the differential as outlined in paragraph 206.

Reverse Idler And Shaft. Reverse idler gear and shaft can be removed after removing transmission cover and detaching transmission housing from torque tube. Bump shaft forward and out of transmission housing wall so that the gear can be removed.

206. R&R TRANSMISSION. Overhaul of the complete transmission requires the following: Drain transmission and differential compartments, and accessory units housing. Block up and support transmission housing and torque tube separately. Disconnect left fender tail light wire at the instru-

Fig. AC90—Model CA transmission shafts, gears and related parts—exploded view. Shims (2) control adjustment of main shaft bearings. Shims (39) control adjustment of bevel pinion shaft bearings. Refer to Fig. AC92 for legend to call-outs 1 through 42.

B. Bevel pinion shaft 1st, 2nd, 3rd, & 4th gear bushings.	56. Snap ring
	57. Spacer (short)
P. Bushing pin	59. Splined coupling collar
39. Shims	60. Spacer
	61. Reverse gear splined collar
	62. Bushing (not available for service)
	63. Reverse idler gear
	64. Idler gear shaft
	65. Idler shaft lock

ment box, and left brake linkage from pedal. Disconnect right brake linkage from pedal by loosening left brake pedal-to-pedal shaft set screw, and moving pedals toward left side of tractor. Disconnect and remove brake pedal-to-brake band toggle linkage.

Remove tractor seat, seat bar, and shock absorber assembly.

Remove both fenders, rockshaft and supports, hydraulic lift control quadrant, platform, and both rear wheel and tire units. Remove hydraulic unit hold positioning valve, rams, and hoses

as an assembly, by removing **two bolts** retaining hold positioning valve to hydraulic pump. Disconnect drawbar to hydraulic lift pump linkage. Loosen drawbar rear bracket retaining bolts to provide clearance for removal of PTO (accessory units) housing. Remove PTO (accessory units) and housing as an assembly. Unbolt and remove both final drive assemblies from transmission housing. Remove cap screws from differential bearing carriers on each side. Pull carriers away from transmission housing; then lift out the bevel ring gear and differential unit rearwards. Remove four nuts retaining transmission housing to torque tube; separate the two assemblies slightly; then remove clutch shaft to transmission mainshaft retaining cotter key. Move torque tube away from transmission.

OVERHAUL. Data on overhauling the various components which make up the transmission are outlined in the following paragraphs. Particular care and attention should be given to the reinstallation of the bevel pinion shaft and gears assembly so as to pre-

Fig. AC91 — Model CA transmission shifter rails, forks, and cover assembly.

45. 1st & 4th gear fork
46. 2nd & 3rd gear fork
48. Reverse gear fork
50. Reverse selector
51. Reverse gear rail
52. 2nd & 3rd gear rail
53. 1st & 4th gear rail

Fig. AC92—Model CA transmission shafts, bevel pinion, ring gear, and differential unit are contained in the transmission housing.

1. Bearing retainer	10. Bearing	19. Shims	26. Bevel ring gear &	34. Gasket
2. Shims	11. Snap ring	20. PTO shaft	differential unit	35. Bearing
3. Bearing	12. Oiling cup	21. Belt pulley shaft	27. 1st gear (50T.)	36. Retaining nut
4. 2nd gear (26T.)	13. Accessory units drive gear	22. Shims	28. Coupling	37. Bevel pinion shaft cover
5. Reverse gear (19T.)	shift lever	23. Accessory units driven &	29. 4th gear (24T.)	38. Bevel pinon shaft
6. 3rd gear (30T.)	14. Hydraulic lift pump	belt pulley drive gear	30. 3rd gear (37T.)	39. Shims
7. 4th gear (49T.)	15. Accessory units drive gear	24. Shims	31. Reverse gear (30T.)	40. Thrust washer
8. Spacer (long)	16. Snap ring	25. Differential unit & bull	32. 2nd gear (41T.)	41. Main shaft
9. 1st gear (18T.)	18. Snap ring	pinion shaft	33. Snap ring	42. Oil seal

vent the loss of the first, second, third, and fourth gear bushing pins.

207. SHIFTER RAILS. Rails and forks can be removed after removing transmission cover. Complete disassembly of rails and forks is self-evident after an examination and reference to Fig. AC91.

208. MAIN SHAFT. Transmission main shaft (41—Fig. AC90 or 92) is removed by first removing transmission from tractor and PTO as described in paragraph 206. Remove PTO sliding drive gear (15) from rear end of main shaft. Remove bearing retainer (1) and bearing adjusting shims from front face of transmission housing.

Bump main shaft forward and out of gears and transmission housing. Front bearing cone (3) will remain on transmission shaft. Center bearing cup (10) will remain in transmission housing, and same can be removed by bumping it in a forward direction. Main shaft rear bearing outer race and roller assembly (17) will remain in the PTO (accessory) housing.

Reassembly is reverse of disassembly. Install second speed gear, 26 *teeth* (4) with flat side rearward; next, install short spacer (57); reverse gear, 19 *teeth* (5) with long part of hub facing rearward; third gear, 30 *teeth* (6) with long part of hub facing forward; fourth gear 49 *teeth* (7) with long part of hub facing rearward; long spacer (8); and low gear, 18 *teeth* (9). Install oil seal (42), which is located in front bearing retainer, with lip of same facing the differential unit.

Shims (2) located between bearing retainer and front face of transmission housing control main shaft bearing adjustment. Adjust to permit shaft to rotate freely but without end play.

209. BEVEL PINION SHAFT. The bevel pinion shaft and gears can be removed by first removing transmission and differential as described in paragraph 206. The bevel pinion shaft can be removed *before* or *after* removing the transmission main shaft. Because of the limited working space provided by the transmission cover opening for reinstallation of pinion shaft, it is advisable to also remove the reverse idler gear and shaft.

Remove bearing cover (37) from front face of transmission housing, and unstake retaining nut (36). Remove retaining nut, and bump bevel pinion shaft rearward and out of gears and transmission housing.

Bevel pinion can be purchased separately from bevel ring gear. Bevel pinion mesh (fore and aft position) is fixed and non-adjustable.

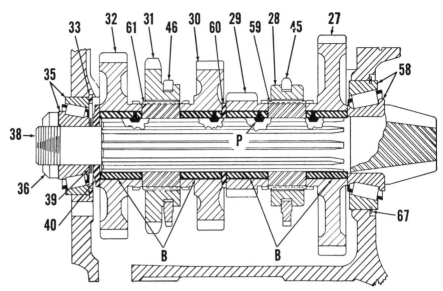

Fig. AC93—Model CA transmission bevel pinion shaft, gears, and related parts assembly. Shims (39) control adjustment of bearings.

B. Bevel pinion shaft 1st, 2nd, 3rd, & 4th gear bushings	31. Reverse gear (30T.)	45. 1st & 4th gear fork
P. Bushing pin	32. 2nd gear (41T.)	46. 2nd & 3rd gear fork
27. 1st gear (50T.)	33. Snap ring	58. Bearing
28. Coupling	35. Bearing	59. Splined coupling collar
29. 4th gear (24T.)	36. Retaining nut	60. Spacer
30. 3rd gear (37T.)	37. Bevel pinion shaft	61. Reverse gear splined collar
	39. Shims	67. Snap ring
	40. Thrust washer	

210. Whenever the bevel pinion shaft, gears, and/or bushings are renewed, it will be advisable to first make a trial assembly on the bench of all the parts to check for free rotation of gears (32, 30, 29, & 27—Fig. AC93). Tighten retaining nuts to 50 ft. lbs. If any gear fails to freely rotate when bevel pinion shaft is held stationary, it is an indication that the gear hub is wider than the bushing (B) on which it rotates. To provide gear side clearance for free rotation, use a fine grade of emery cloth on a flat surface to decrease the width (flat sides) of the gear.

211. Reassembly is reverse of disassembly. As an aid in reassembling, remove front bearing cup (35) from transmission housing.

CAUTION: The low, second, third, and fourth speed gears rotate on hardened steel bushings (B). Each bushing is equipped with a small, removable steel pin (P) which engages a spline in the bevel pinion shaft. Renew pins if damaged in any way and rivet the pins in the bushings to prevent their falling out during the reassembly of the pinion shaft.

Reinstall bevel pinion shaft in housing, and working through transmission cover opening, install gears, thrust washers, and shims, as shown in Fig. AC93.

211A. *Bearing Adjustment*. To adjust bevel pinion shaft bearings, add several more shims (39) to the number originally removed so as to provide

Fig. AC94—Model CA bevel pinion shaft 2nd gear (41T.), steel bushing and pin assembly. The bevel pinion shaft 1st, 3rd, and 4th gears rotate on similar bushings.

shaft with end play. Shims are available in .003, .005, and .010 thickness. Next, install bearing cup and cone, and tighten retaining nut to 50 ft. lbs. torque. With a dial indicator, measure and record pinion shaft end play then remove shims equal in amount to the recorded pinion shaft end play so that shaft rotates freely, yet without perceptible end play. If the front bearing cone is a tight fit on the shaft it will be advisable in removing the shims to remove the bevel pinion shaft and gears. After bearings are adjusted, tighten nut (36) to 50 ft. lbs., and lock same in position by staking.

After installing differential unit, adjust carrier bearings by means of

Fig. AC95—Model CA reverse idler gear (63) and shaft (64) installation viewed through transmission cover opening.

shims (38—Fig. AC109) so that the unit rotates freely, yet without perceptible end play.

211B. *Backlash Adjustment.* After adjusting end play adjust bevel pinion-to-ring gear backlash to .005-.014 by removing shims from under one differential bearing carrier and installing them under the opposite bearing carrier. To increase backlash remove shim from ring gear side and install the same shim to carrier on opposite side. This adjustment should be made *after*, never *before*, the bearings have been adjusted as in preceding paragraph.

212. **REVERSE IDLER AND SHAFT.** Reverse idler gear (63—Fig. AC95) and shaft (64) can be removed whenever the transmission is detached from the torque tube by bumping the shaft out forward, while working through transmission cover opening.

The reverse idler gear rotates on a bronze bushing. Worn bushings are serviced by renewing the gear and bushing as an assembly. Install reverse gear with shifter collar facing rearward.

Model G

213. **BASIC PROCEDURES.** Transmission shafts, differential, and the final drive bevel gears are carried in one case. Although most transmission repair jobs involve overhaul of the complete unit, there are infrequent instances where the failed or worn part is so located that the repair work can be completed safely, without complete disassembly of the transmission. In effecting such localized repairs, the mechanic will save time by observing the following as a general guide: Refer to Fig. AC96.

A. On 3 speed transmissions the shifter rails and forks are accessible for overhaul after removing only the battery box and the control (transmission) cover. On 4 speed transmissions the renewal or overhaul of the extra fork and shaft of the extra low gear requires detaching transmission housing from main frame.

B. Main shaft bearings and bevel pinion shaft bearings can be adjusted

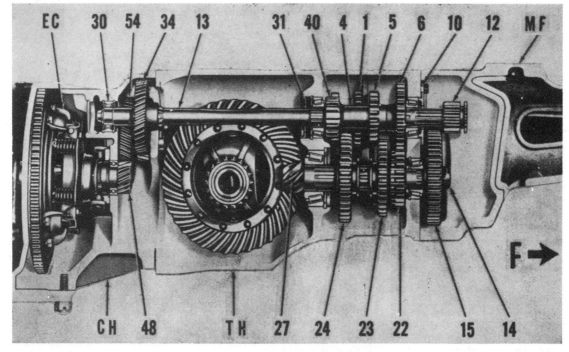

Fig. AC96—Model G transmission shafts, bevel pinion, ring gear, and differential unit are contained in the transmission housing.

CH. Clutch housing	5. 2nd gear	15. Extra low gear	31. Oil shield
EC. Engine clutch	6. 3rd gear	22. 3rd gear	34. Main intermediate gear
MF. Tractor main frame	10. Shims	23. 2nd gear	40. 1st reverse gear
TH. Transmission housing	12. Extra low pinion	24. 1st & reverse gear	48. Clutch shaft & gear
1. Reverse idler	13. Mainshaft	27. Bevel pinion	54. Idler & belt pulley
4. Spacer	14. Adjusting nut	30. Roller bearing	drive gear

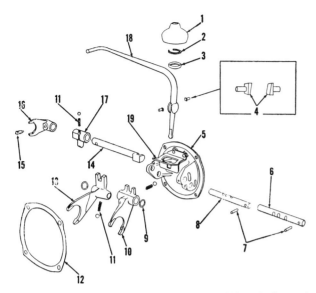

Fig. AC97—Model G transmission cover, shifter forks and rails.

1. Cap	10. 1st & reverse fork
2. Snap ring	11. Spring
3. Retainer	12. Gasket
4. Lever guide pins	13. 2nd & 3rd fork
5. Transmission cover	14. Extra low rail
6. 1st & reverse rail	15. Retaining screw
7. Retaining pins	16. Extra low fork
8. 2nd & 3rd rail	17. Rail guide
9. Overshift washers	19. Lever guide

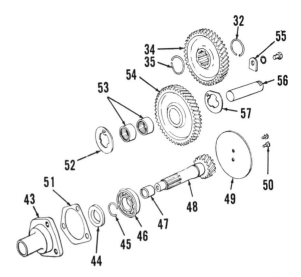

Fig. AC98—Model G clutch shaft, and transmission intermediate gears. All of these parts are contained in the clutch housing.

32. Snap ring	49. Oil slinger
34. Mainshaft intermediate	51. Gasket
gear	52. Thrust washer (rear)
35. Snap ring	53. Roller bearings
43. Bearing retainer	54. Idler & belt pulley drive
44. Oil seal	gear
46. Clutch shaft bearing	55. Idler shaft lock
47. Clutch shaft pilot bushing	56. Idler gear shaft
48. Clutch shaft & gear	57. Thrust washer (front)

after detaching transmission housing from main frame.

C. On 4 speed transmissions the extra sliding and stationary low speed gears can be renewed after detaching transmission housing from main frame.

D. The transmission main shaft intermediate gear (34) which is located in the clutch housing can be renewed after detaching the clutch housing from the transmission case.

E. Removal of the clutch shaft (48) requires removal of the clutch housing which also necessitates removal of the engine from the tractor.

F. Renewal or overhaul of the mainshaft (13) requires detaching clutch housing from the transmission case and the transmission from the main frame.

G. Renewal or overhaul of the bevel pinion shaft (27) requires removal of the main shaft and differential, which in turn necessitates removal of both rear axle shaft housing units.

R & R AND OVERHAUL. Data on overhauling the various components of this model are outlined below.

214. SHIFTER FORKS. On tractors equipped with 3 speed transmissions, these parts can be renewed after removing the battery box and control cover. The gear shift lever guide pins (4) shown in Fig. AC97 are located in the transmission case and have beveled heads as shown. On 3 speed transmissions the right hand guide pin (viewed from control cover side of

transmission) should be installed with the thick side of the head up; the left hand pin is positioned with the thick side down. On 4 speed transmissions the pins should be installed by reversing the foregoing arrangement. The extra shift fork and shaft on these models is located in the transmission and is not accessible for renewal until the main frame is disconnected, as outlined in a preceding paragraph.

215. CLUTCH SHAFT. Removal of shaft necessitates removal of the engine as outlined under R & R ENGINE. After the engine is removed disconnect fuel line from tank and remove the radiator. Remove belt pulley from clutch housing if tractor is so equipped. Remove clutch actuating rod and clutch housing to transmission case bolts and the clutch housing. Contained in the housing are the clutch shaft (48—Fig. AC98), release fork and belt pulley drive gear (54). Remove oil slinger (49) from gear end of shaft and the bearing retainer (43) from the front end. The clutch shaft can now be bumped out of the housing in the direction towards the flywheel. The lip of the oil seal (44) on clutch shaft should face toward the transmission.

Idler and belt pulley drive gear can now be removed by removing shaft (56) toward transmission. Front thrust washer (57) has a flattened edge and should be installed on transmission side of gear (54).

216. MAIN INTERMEDIATE GEAR. When the transmission is disconnected from the clutch housing the intermediate gear (34) is accessible for removal or renewal. Gear is retained to shaft by snap rings. The bearing (30—Fig. AC99), for the rear end of the mainshaft, which is located in the clutch housing, can now be renewed.

217. MAIN FRAME. Procedure for removing this part, which at its front end is bolted to the tubular steel frame and at its rear end to the transmission case, is readily determined after referring to the unit. When reconnecting the main frame (cast frame) be sure that the sealing washers of copper are in place under the two lower nuts where the tube bolts to the transmission case.

218. EXTRA LOW GEARS. On 4 speed gearsets, the extra low sliding pinion (12—Fig. AC96 or 99), which is movably splined on the forward end of the main shaft, and the stationary gear (15), which is retained on the forward end of the bevel pinion shaft, can be renewed independently as follows:

A. Drain transmission case and rear portion of main frame, then disconnect and separate the main frame from the front of the transmission case.

B. Remove the bevel pinion shaft nut (14), and pull the stationary gear from the shaft. When reinstalling gear, tighten nut until all bearing play is removed and a slight preload exists.

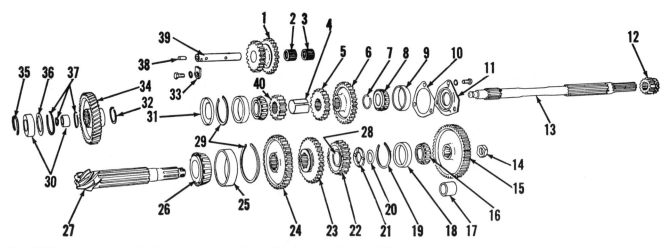

Fig. AC99—Arrangement of shafts and gears in model G transmission. Shims (10) control mainshaft bearings adjustment and nut (14) controls bevel pinion shaft bearings adjustment. Bushing (28) in third speed gear can be purchased separately.

1. Reverse idler	11. Bearing retainer	20. Belville spring washer	27. Bevel pinion	34. Mainshaft gear
2. Idler bearing	12. Extra low pinion	21. Thrust washer	28. Bushings	35. Snap ring
4. Spacer	13. Mainshaft	22. 3rd gear	29. Snap rings	36. Oil shield
5. 2nd gear	14. Adjusting nut	23. 2nd gear	30. Roller bearing	37. Snap rings
6. 3rd gear	15. Extra low gear	24. 1st & reverse gear	31. Oil shield	38. Idler shaft plug
7. Snap ring	17. Spacer (3 speed only)	25. Bearing cup	32. Snap ring	39. Idler shaft
10. Shims	19. Snap ring	26. Bearing cone	33. Idler shaft lock	40. 1st & reverse gear

Lock the nut by driving a portion of same into the shaft keyway.

C. Remove the sliding pinion (12) by removing the set screw from shifter fork which will release the fork and the pinion.

219. ADJUST BEVEL PINION BEARINGS. To adjust bearings, follow the procedure outlined in paragraph 218B, but do not remove the gear on 4 speed units or the spacer (17) on 3 speed units.

220. MAIN SHAFT. To remove and overhaul shaft, proceed as follows: Remove the transmission case from tractor by disconnecting same from clutch housing and torque tube or main frame. Remove control cover unit from side of transmission case. It is not necessary at this time to remove the differential unit. Remove main shaft front cap or bearing retainer (11—Fig. AC99), and the intermediate gear (34) from opposite end of shaft. Bump shaft forward out of case while removing gears and spacers through control (shift lever) cover opening. After renewing worn parts adjust bearings to zero end play and a free rolling fit by adding or removing shims (10) interposed beween case and front bearing cap or retainer.

221. REVERSE IDLER GEAR. First step in overhauling these parts is to remove the main shaft as in paragraph 220. Remove idler lock (33) from outside of case and pull shaft rearward. Outer end of shaft is threaded internally for a puller. The cork plug (38) in the rear end of shaft acts as an oil seal. Be sure it is in good condition.

222. BEVEL PINION SHAFT. First step in overhauling shaft is to remove the engine, the main shaft, and the differential assembly which necessitates removal of the rear axle housings. Remove pinion shaft nut (14) and bump shaft rearward out of case. Remove gears, thrust washer, and Belville spring washer (20) out through cover opening. If the pinion is to be renewed for any reason, it will be necessary to purchase the bevel ring gear also as they are furnished only as a matched set. Pinion mesh position is fixed and non-adjustable.

When reinstalling the parts, assemble the sliding gears (23) and (24), as shown with their shift collars facing each other. Install the 3rd speed gear (22) as shown. The thrust washer (21) for this gear should be installed with the grooves facing the 3rd speed gear and plain side facing the Belville spring washer (20), as shown. If the transmission is a 3 speed unit, install the spacer (17) instead of the special low gear (15), which is used only on 4 speed units.

After installing nut (14) tighten same to adjust bearings to a just perceptible preload. If only the bevel pinion shaft is in the case, the correct preload is when ½ lb. of torque measured at end of a one-ft. wrench or arm is required to rotate the shaft in its bearings. After adjustment is completed, lock the nut by staking thin section of same into shaft keyway.

Models RC-WC-WF

223. R & R TRANSMISSION UNIT. Support tractor under frame and detach rear axle housing from frame channels. Remove steering support brace and fender rods from front of fenders and frame channels. (On WF model, remove platform and steering gear.) Roll rear of tractor back, withdrawing drive shaft and torque tube from the transmission mainshaft sleeve and rear bearing retaining cap. Remove inspection cover from bottom of clutch housing and disconnect clutch shifter fork spring from transmission stud and remove nut from stud. Remove cap screws retaining transmission to clutch housing and pull transmission rearward.

224. CONTROL COVER can be removed from transmission after removing cap screws retaining cover to transmission case.

225. SHIFTER RODS AND FORKS can be disassembled after removing transmission cover. Turn up lip of gear shift lever dust cap and remove snap ring (4—Fig. AC101), washer (5), and gear shift lever. Loosen shifter fork screws and lug screws and drive expansion plugs out of front of cover. Push shifter rods (13, 17, and 18) out through front of cover and remove forks from bottom of cover being careful not to lose detent balls (7) and springs (6) from holes in cover and interlock plugs (22) and pin (23) from center shifter rod.

OVERHAUL. Procedures for overhauling the various parts of the transmission unit are contained in the following paragraphs:

226. CLUTCH SHAFT. Clutch shaft (1—Fig. AC100) is positioned by snap ring (4) on main drive gear bearing outer race and can be pulled from the case after separating transmission from the clutch housing as in paragraph 223. To remove belt pulley bevel pinion gear (3) and main drive gear bearing, remove snap ring (2) and press pinion gear and bearing off shaft.

Cork seals (7—Fig. AC75) for clutch shaft which are located in clutch housing can be renewed at this time.

Reassembly is reverse of disassembly. Be sure spigot bearing (mainshaft pilot bearing) (20—Fig. AC100) is in position on mainshaft.

227. MAIN (SPLINE) SHAFT. Removal of this shaft (8) requires removing transmission cover and disconnecting rear of tractor from transmission case as described in paragraph 223. Remove transmission rear bearing cap (10) and pull mainshaft (8), bearing and sleeve (12) out through the rear of case while withdrawing sliding gears from top of case. Spigot bearing (mainshaft pilot bearing) (20) can be removed from pocket in clutch shaft. Mainshaft rear bearing can be pressed off shaft after driving riveted pin (13) out of sleeve and removing sleeve.

Reassembly is reverse of disassembly. Install rear bearing and sleeve on shaft and rivet sleeve retaining pin. Be sure spigot bearing and spigot bearing spacer (19) are in position and insert shaft through rear of case and install gears on shaft as follows: First, second, and reverse sliding gear (7) with groove toward front; third and fourth speed sliding gear (6) with groove toward rear.

228. COUNTERSHAFT. To remove this shaft, it is first necessary to remove the clutch shaft, and mainshaft. Then remove cap screw and lock plate (21) from rear of case and bump

countershaft out toward the rear and withdraw gear cluster (17) through the top of the case.

229. REVERSE IDLER SHAFT. To remove this shaft it is first necessary

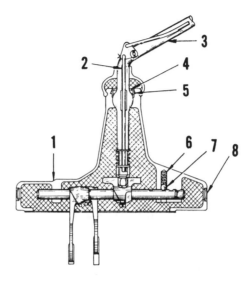

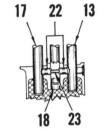

Fig. AC101—Models RC, WC, and WF transmission control and cover assembly—side and rear views.

2. Reverse latch rail	13. 1st & 2nd shift rail
3. Shift lever	14. 1st & 2nd shift fork
4. Snap ring	15. 3rd & 4th shift fork
5. Shift lever washer	16. Reverse shift fork
6. Detent ball spring	17. Reverse shift rail
7. Detent ball	18. 3rd & 4th shift rail
8. Expansion plug	19. Reverse shift lug
9. Reverse latch spring	20. Shift lever pivot pin
10. Reverse latch	21. Dust cap
11. 3rd & 4th shift lug	22. Shift rail interlock
12. 1st & 2nd shift lug	23. Shift rail interlock pin

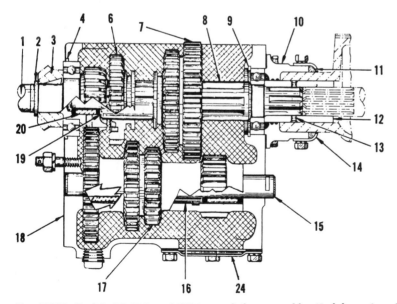

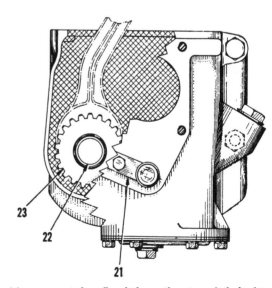

Fig. AC100—Models RC, WC, and WF transmission assembly. Models equipped with a power take-off unit have the pto unit bolted to the lower side of the transmission housing, at "24". Drive for the pto is supplied through a countershaft gear.

1. Clutch shaft	6. 3rd & 4th sliding gear	11. Mainshaft sleeve packing	16. Countershaft bearing	20. Spigot bearing
2. Snap ring	7. 1st & 2nd sliding gear	12. Mainshaft sleeve	spacer	21. Countershaft and idler
3. Belt pulley bevel driving	8. Main (spline) shaft	13. Mainshaft sleeve pin	17. Countershaft gear cluster	gear shaft lock plate
gear	9. Snap ring	14. Packing retainer	18. Transmission case	22. Idler gear shaft
4. Front bearing snap ring	10. Rear bearing cap	15. Countershaft	19. Spigot bearing spacer	23. Reverse idler gear

to remove transmission and mainshaft as in paragraph 227. Remove lock plate (21) from rear of case and bump idler gear shaft out toward the rear and withdraw gear through the top of the case. Bushing in reverse gear is renewable and must be reamed after installation, to provide .001-.002 inch clearance.

Model WD Prior 127008

First type transmissions have a straight (not curved) gear shifting lever and straight cut gear teeth.

230. **BASIC PROCEDURE.** Transmission shafts, differential unit and final drive bevel pinion and ring gear are carried in the one housing. Overhaul of any of the transmission shafts, except the main drive shaft (transmission clutch shaft), requires removal of transmission as outlined in paragraph 231.

231. **R & R TRANSMISSION.** Drain oil from torque tube, transmission and differential housing. Support tractor under torque tube and rear portion of main frame and transmission. Remove rear wheels and final drive assemblies as outlined in F I N A L DRIVE section, and differential and bevel ring gear unit as outlined in DIFFERENTIAL section. Remove drawbar, battery, battery carrier and power take-off shaft. Remove steering column to transmission housing retaining cap screws. Disconnect brake pedal rods and engage transmission clutch. Remove rear platform and power lift arms and shaft. Remove main frame to transmission retaining screws. Remove nuts retaining transmission to torque tube. Roll transmission housing away from torque tube. Refer to Fig. AC102.

OVERHAUL. Data on overhauling the various components of this model are outlined below:

232. **SHIFTER R A I L S , FORKS.** Shifter forks and rails are contained in the transmission top cover, as shown in Fig. AC103. Transmission cover can be removed after removing battery, battery carrier and cover retaining cap screws. Method of checking and overhauling the shifter rails and forks is conventional.

233. **MAIN DRIVE SHAFT.** This is also the transmission clutch shaft (7—Fig. AC104) or input shaft. To remove and overhaul this shaft, proceed as follows: Detach transmission housing from torque tube and roll transmission and final drive assembly away from tractor. Remove transmission top cover. Working through top cover open-

Fig. AC102—Model WD showing use of special long pilot studs which facilitate detaching the transmission and final drive assembly from the torque tube and frame rails. Model WD45 is similar.

ing, bump gear end of main drive shaft forward and out of transmission housing.

Transmission bevel pinion shaft needle roller pilot bearing (2) is contained in gear end of main drive shaft and can be renewed at this time. Transmission main drive pilot bushing (B—AC84) (located in rear end of engine clutch shaft) should be renewed at this time if clearance exceeds .005.

234. **BEVEL PINION.** Remove transmission main drive shaft as outlined in preceding paragraph, and remove the differential and bevel ring gear unit as outlined in DIFFERENTIAL section. Working through the differential compartment, remove bevel pinion shaft bearing retainer (1—Fig. AC104). After removing the 6 retaining cap screws bump (or preferably push with a portable hydraulic jack) bevel pinion shaft, together with bearing cup and cones (3), rearward and remove the first, second and reverse, and third and fourth speed gears out through top of housing.

Bevel pinion and shaft can be purchased separately from the mating bevel ring gear. Bevel pinion mesh is fixed and non-adjustable.

Assemble pinion shaft and bearings on the bench and adjust bearings by means of nut (4) to a just perceptible preload of 10-20 inch pounds. In making up the assembly, install the wider roller bearing (one with longer rollers) next to the bevel pinion.

To facilitate reinstallation of pinion and bearings assembly to transmission case freeze same in dry ice or freeze chest for at least half an hour.

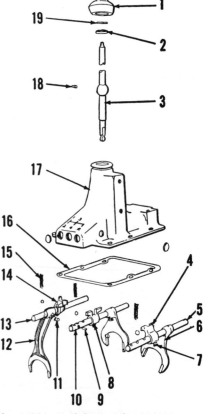

Fig. AC103—Model WD prior 127008 transmission cover, shifter forks and rails as used on early production (first type) transmissions. To correct an over-shift condition Allis-Chalmers supply a sleeve type spacer to be installed on the forward end of reverse rail (13).

2. Retainer	10. 1st & 2nd rail
4. 3rd & 4th selector	11. Interlock plunger
5. 3rd & 4th rail	12. Reverse fork
6. 3rd & 4th fork	14. Reverse selector
7. Set screw	15. Detent spring
8. 1st & 2nd selector	18. Pivot pin
9. Interlock pin	19. Snap ring

1. Bearing retainer
2. Bevel pinion shaft pocket (pilot) bearing
4. Adjusting nut
5. 1st, 2nd & reverse gear
6. 3rd & 4th gear
7. Main drive gear & shaft
8. Bevel pinion & shaft
11. Countershaft gear cluster
12. Bearing
13. Reverse idler gear
14. Snap ring
15. Idler gear bushing
16. Snap ring
17. Spacer
18. Roller bearing
19. Countershaft
20. Countershaft & reverse shaft lock
21. Reverse idler shaft

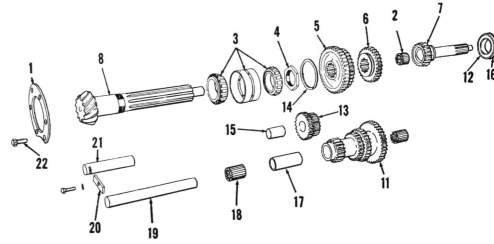

Fig. AC104—Arrangements of shafts and gears in early Model WD transmission. Nut (4) controls bevel pinion shaft bearing adjustment.

235. COUNTERSHAFT. To remove the countershaft gears (11), first remove the main drive shaft and bevel pinion shaft as outlined in preceding paragraphs. Working through the differential compartment, remove lock plate (20). Bump countershaft (19) rearward and out of transmission case. Cluster gear rotates on two roller bearings.

236. REVERSE IDLER GEAR. To remove and overhaul the reverse idler and shaft, remove the main drive shaft and bevel pinion shaft as outlined in preceding paragraphs. It is not necessary to remove the countershaft. Working through the differential compartment, remove lock plate (20) and bump reverse idler shaft rearward and out of transmission case. Reverse idler (13) rotates on a bronze bushing (15) which is renewable.

MODEL WD45

Models WD after 127007 & WD45

This unit has a curved lever and helical cut gears.

237. BASIC PROCEDURE. Transmission shafts, differential unit and final drive bevel pinion and ring gear are carried in the one housing. Overhaul of any of the transmission shafts, except the main drive shaft (transmission clutch or input shaft), requires removal of transmission as outlined in paragraph 231.

238. R&R TRANSMISSION. Follow procedure as outlined for model WD prior 127008 transmission in paragraph 231.

OVERHAUL. Data on overhauling the various components of this model are outlined in the following paragraphs.

239. SHIFTER RAILS, FORKS. Shifter forks and rails are contained in the transmission top cover as shown in Fig. AC105B. Transmission cover can be removed after removing battery, battery carrier and cover retaining cap screws. Method of checking and overhauling the shifter forks and rails is conventional.

239A. MAIN DRIVE SHAFT. This is also the transmission clutch shaft or input shaft (23—Fig. AC105A). To remove and overhaul this shaft, proceed as outlined for model WD prior 127008 transmission in paragraph 233.

239B. BEVEL PINION. Bevel pinion shaft and gear is removed **before** removing the countershaft. Remove transmission assembly from the tractor as outlined in paragraph 231, and

bevel ring gear and differential assembly as outlined in paragraph 251. Remove main drive shaft as outlined in paragraph 233, and shifter cover assembly from top of transmission. Working through the differential compartment, remove 6 cap screws retaining bevel pinion shaft bearing retainer (3) to transmission housing. Working through transmission housing top cover opening, remove snap ring (21) from forward end of bevel pinion shaft. Bump (or preferably push with a portable hydraulic jack) bevel pinion shaft rearward, together with bearing cup and cones (4), and remove gears, bushings, washer and thrust washer out through top of housing.

Caution: The first, second, and third gears rotate on hardened steel bushings (11A, 11B & 17). Each bushing is

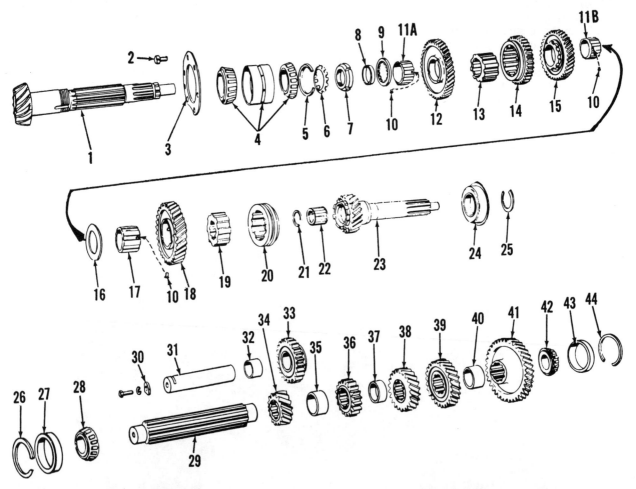

Fig. AC105A—Arrangement of shafts, gears and related parts in model WD after 127007 and WD45 transmission (second type). Nut (7) controls bevel pinion shaft bearings adjustment. Thickness of snap ring (44) controls countershaft bearings adjustment. Snap rings 5, 21 and 44) are available in various thicknesses. Model WD45 is similar.

1. Bevel pinion & shaft	11A. Bushing (large)	18. 3rd gear (31 teeth)
2. Cap screw	11B. Bushing (large)	19. 3rd & 4th gear splined
3. Bearing retainer	12. 1st gear (40 teeth)	collar
4. Bearing cup & cones	13. 1st & 2nd gear splined	20. 3rd & 4th gear splined
5. Snap ring	collar	coupling
6. Lock washer	14. 2nd & reverse sliding	21. Snap ring
7. Adjusting nut	gear (34 teeth)	22. Pinion shaft pilot bearing
8. Snap ring	15. 2nd gear (35 teeth)	23. Main drive (4th) gear &
9. Washer	16. Thrust washer	shaft
10. Bushing pin	17. Bushing (small)	24. Bearing & snap ring

25. Snap ring	35. Spacer (.966 wide)	
26. Snap ring	36. Reverse gear (18 teeth)	
27. Bearing cup (rear)	37. Spacer (.654 wide)	
28. Bearing cone (rear)	38. 2nd gear (22 teeth)	
29. Countershaft	39. 3rd gear (26 teeth)	
30. Reverse shaft lock	40. Spacer	
31. Reverse shaft	41. Drive gear (37 teeth)	
32. Reverse gear bushing	42. Bearing cone (front)	
33. Reverse gear	43. Bearing cup (front)	
34. 1st gear (16 teeth)	44. Snap ring	

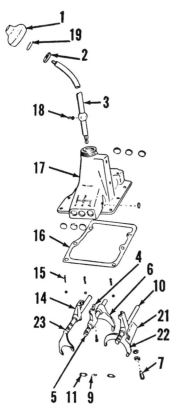

Fig. AC105B—Model WD transmission cover, shifter forks and rails as installed on the synchro-mesh transmissions, Fig. AC-105A. Model WD45 is similar.

1. Dust cover
2. Retainer
3. Shift lever
4. 3rd & 4th selector
5. 3rd & 4th rail
6. 3rd & 4th fork
7. Set screw
9. Interlock pin
10. 1st & 2nd rail
11. Interlock plunger
14. Reverse selector
15. Detent spring
16. Gasket
18. Pivot pin
19. Snap ring
21. Oil shield
22. 1st & second fork & selector
23. Reverse rail & fork

equipped with a small, removable steel pin (10) which engages a spline on the bevel pinion shaft.

Bevel pinion can be purchased separately from bevel ring gear. Bevel pinion mesh (fore and aft position) is fixed and non-adjustable. However, snap ring (5), which is available in several thicknesses, controls the cone center distance for a particular transmission housing. If necessary to renew the ring be sure that replacement is of exactly the same thickness. This does not apply where a new transmission housing is being installed as it is assumed that the new housing will include a factory fitted snap ring.

Whenever the bevel pinion shaft, gears and/or bushings are renewed, it will be advisable to first make a trial assembly on the bench of all parts located between and including the front and rear snap rings (8 & 21) to check

for free rotation of gears (12, 15 & 18). The front snap ring (21) is available in several thicknesses. Select a snap ring which will eliminate all end play of the assembled parts when same are installed on the pinion shaft. If any gear fails to freely rotate when bevel pinion shaft is held stationary, it is an indication that the gear hub is wider than the bushing on which it rotates. To provide gear side clearance for free rotation, use a fine grade of emery cloth on a flat surface to decrease the width (flat sides) of the gear.

Reassembly is reverse of disassembly. Assemble pinion shaft and bearings on the bench and adjust bearings by means of nut (7) to a just perceptible preload of 10-20 inch pounds and lock the adjustment with one of the tabs on lock washer (6). In making up the assembly, install the wider roller bearing (one with longer rollers) next to the bevel pinion. To facilitate re-installation of pinion and bearings assembly to transmission case, freeze same in dry ice or freeze chest for at least half an hour.

Enter pinion shaft and bearing assembly in transmission housing and install pinion shaft components in the following order: Snap ring (8), washer (9), first gear bushing (11A) with lock pin hole facing forward, first speed gear (12) 40 *teeth* with clutch jaws forward, first and second gear splined collar (13), second and reverse sliding gear (14) 34 *teeth* with shifter fork groove rearward, second gear bushing (11B), second gear (15) 35 *teeth* with clutch jaws rearward, thrust washer (16), third gear bushing (smallest) (17) with lock pin facing forward, third speed gear (18) 31 *teeth* with clutch jaws forward, third and fourth gear splined collar (19), snap ring (21) which is available in several thicknesses, and third and fourth gear splined coupling (20) with shifter fork groove rearward. Install bevel pinion shaft pilot bearing (22). To the front end of transmission housing install main drive shaft assembly. Working through differential compartment, install bevel pinion shaft bearing retainer cap screws (2) and torque same to 35 foot pounds. Using soft, flexible wire, threaded through the drilled heads of the cap screws, lock retainer cap screws in position.

After installing differential unit, adjust carrier bearings by means of shims so that the unit rotates freely, yet without perceptible end play. Af-

ter adjusting the bearings, adjust bevel pinion-to-ring gear backlash to .005-.014 by removing shims from under one differential bearing carrier and installing them under the opposite bearing carrier. To increase the backlash, remove shims from ring gear side and install the same shims under the carrier on the opposite side. This adjustment should be made **after,** never **before,** the bearings have been adjusted.

239C. COUNTERSHAFT. To remove the countershaft and gears, first remove the main drive shaft, and bevel pinion shaft as outlined in preceding paragraphs 239A and 239B. From front of transmission housing, remove countershaft front bearing cup retaining snap ring (44). Snap ring (44), which is available in several thicknesses, is used to adjust the countershaft bearings. Working through differential compartment, bump countershaft forward and out of transmission housing, and remove countershaft gears, rear bearing cone and spacers out through transmission top cover opening.

Whenever the countershaft and/or bearings are renewed, or the countershaft bearings require adjusting, it will be advisable to first make a trial installation of only the countershaft and bearings in the transmission housing so as to adjust the shaft end play. Shaft end play should be adjusted to .003-.005 by varying the thickness of the front bearing cup retaining snap ring (44). After making the bearing adjustment, remove the snap ring bearing cup, shaft, and rear bearing cone.

Reassembly is reverse of disassembly. Install rear bearing cup (27) and snap ring (26) in transmission housing. Install bearing front cone (42) with taper toward end of shaft. Next, from the forward end of the transmission housing, enter the shaft through front bearing bore, and install the gears and spacers in the following order: Drive gear (41) 37 *teeth* with long hub rearward, spacer (40), third gear (39) 26 *teeth* with long hub forward, second gear (38) 22 *teeth* with long hub forward, spacer (37) which is .654 wide, reverse gear (36) 18 *teeth* with bevel of teeth rearward, spacer (35) which is .966 wide. Then, before installing the first gear (34) 16 *teeth*, place the rear bearing cone (28) in the rear bearing cup. Holding the rear bearing cone in this position, install the first gear (34) and push countershaft rearward until rear bearing cone is installed on the shaft. Install front bearing cup (43) and special snap ring (44). Recheck countershaft end play which should be .003-.005.

239D. REVERSE GEAR. To remove and overhaul the reverse gear (33) and shaft (31), first remove the main drive shaft, and bevel pinion shaft as outlined in preceding paragraphs. It is not necessary to remove the countershaft. Working through the differential compartment, remove lock plate (30) and bump reverse shaft rearward and out of transmission housing, and remove gear out through top cover opening. Reverse idler gear rotates on a bronze bushing (32) which is renewable.

MODEL C

MODEL G

DIFFERENTIAL, FINAL DRIVE & REAR AXLE

DIFFERENTIAL

Models B-C-CA

Differential unit is of the two-pinion, open case type, and is located back of the dividing wall cast into the transmission housing. Differential carrier bearings are shim adjusted, tapered roller type, mounted in carriers (35—Fig. AC106), which are bolted to the transmission housing. The carriers are covered by their respective final drive units (12), which must be removed before any work can be done on the carriers.

240. BEARINGS AND BACKLASH ADJUSTMENT. Carrier bearings are adjusted by varying the number and thickness of shims (38) located under the carriers. Although shim removal can be accomplished without removing the accessory (PTO) housing (bolted to the rear face of the transmission), or transmission housing rear cover, there is no sure way of checking the bearing adjustment or the pinion to ring gear backlash without doing so.

To adjust bearings, first remove both final drive assemblies as outlined in paragraph 269 for B and C or 274 for the CA. Remove the PTO and BP unit (or the complete accessory unit of CA) from rear face of transmission housing on models so equipped, or transmission housing rear cover. On CA models with continuous type PTO, remove final drive clutch actuator support (39—Fig. AC128) from right side of transmission housing.

240A. Select shims (38—Fig. AC106) between carriers and housing to remove all bearing play but permitting differential to turn without binding. Removing shims reduces bearing play. Divide shims between left and right carriers to provide .005-.014 backlash between teeth of pinion and ring gear. To increase backlash (after bearings have been adjusted), remove shim or

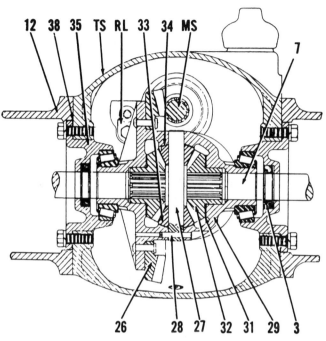

12 38 35 TS RL 33 34 MS

7

MS. Transmission mainshaft
RL. Reverse idler gear shaft lock plate
TS. Transmission housing
3. Shaft oil seal
7. Final drive pinion shaft
12. Final drive housing
26. Main drive bevel ring gear
27. Differential pinion gears shaft
28. Differential pinion shaft lockpin
29. Differential case
31. Thrust washer
32. Differential side gear
33. Differential pinion
34. Pinion gear thrust washer
35. Bearing carrier
38. Bearing shims

26 28 27 32 31 29 3

Fig. AC106—Models B, and C differential and bevel ring gear assembly—rear view. Model CA is similar.

MODEL CA

Fig. AC107—Model CA two pinion, open case type differ-ential unit is located back of the transmission housing dividing wall. Sliding spur gear (36) on transmission mainshaft (MS) supplies power to the accessory units. Models B, and C are similar.

MS. Transmission mainshaft
35. Differential bearing carrier
36. Sliding gear
37. Snap ring

is originally riveted to the differential case. The preferred method of re-moving the rivets is by drilling. If ring gear is removed, it may be re-attached with cold rivets or (on B, C, and CA only) by use of heat-treated alloy bolts available from Allis-Chalmers dealers. Don't use ordinary bolts. When re-riveting, temporarily bolt the gear to the differential case, and be careful to avoid distortion of gear or case. After ring gear is attached, check trueness at ring gear back face with a dial indicator with unit in its car-riers or between centers of a lathe. Total run-out should not exceed .003.

After unit is reinstalled in transmis-sion, adjust bearings and backlash as outlined in paragraph 240A.

shims from carrier on ring gear side of housing and install same under carrier on opposite side.

241. R & R AND OVERHAUL. To remove the differential unit from the transmission housing, first drain the transmission and pto or accessory unit housings. Remove both final drive units as outlined in paragraph 269 for B and C or 274 for the CA. Remove the pto main housing from rear face of transmission. Remove differential carrier (each held to transmission by four cap screws) from each side of transmission, including the final drive actuator clutch support (39—Fig. AC128) on CA. Do not mix the shims. Tilt differential down and toward the left and lift it from the transmission.

241A. To disassemble differential, drive out the lock pin (28—Fig. AC106) (drive from left to right) and remove pinion pin (27). Pinion gears (33), side gears (32), and thrust wash-ers (31 & 34) can then be removed from the case.

Recommended backlash of .005 be-tween teeth of side gears and teeth of pinions is controlled by the copper plated side gear thrust washers (31) and/or the pinion thrust washers (34). Pinion gears are equipped with bush-ings, but the bushings are not avail-able for service.

The bevel ring gear, which is avail-able separately from the bevel pinion,

26. Bevel ring gear
27. Pinion shaft
28. Pinion shaft retaining pin
29. Differential case
31. Thrust washers
32. Side gears
33. Pinion gears
34. Thrust washers

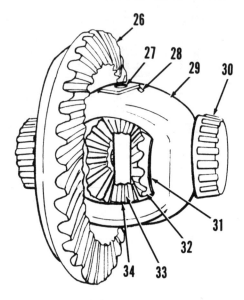

Fig. AC108—Models B, C, and CA main drive bevel ring gear, and differential unit. Bevel ring gear is available separately from the bevel pinion. Either cold rivets or special heat-treated steel bolts and nuts for attaching the ring gear to the case are supplied for service.

3. Oil seal (in carrier)
35. Differential bearing carrier
38. Shims

Fig. AC109—Model CA differential bearing carriers are acces-sible after removing the final drive housing. Models B, and C are similar. On CA models equipped with continuous type PTO, also remove the final drive clutch actuator support from the right side.

Model G

The main drive bevel pinion and ring gear are available only as a matched set; neither can be purchased separately.

242. BEARINGS AND BACKLASH ADJUSTMENT. Carrier bearings are adjusted by varying the number and thickness of shims (11—Fig. AC110) located under each carrier. Although shim removal can be accomplished without detaching the engine clutch housing from the transmission, there is no sure way of checking the bearing adjustment or the pinion to ring gear backlash without doing so.

To adjust bearings, it is necessary to first remove both axle shaft and sleeve units as outlined in paragraph 286. Remove each backing plate and its brake shoes as an assembly. Detach engine and clutch housing from transmission housing. Remove bearing carriers (10) then adjust bearings to zero end play and bevel pinion to ring gear backlash to .005-.007 as outlined in paragraph 240A.

243. R & R AND OVERHAUL. Main drive bevel ring gear and differential unit are carried in the transmission housing rear compartment. To remove the differential unit, proceed as for adjusting bearings as outlined in paragraph 242 then lift out the differential.

244. To overhaul the two pinion, open case type differential unit on the bench proceed as outlined in paragraph 241A.

Fig. AC111—Model RC bevel ring gear and differential unit assembly—rear view. (B) Bolts retaining rear axle housing to tractor frame rails. (S) Nuts & studs retaining differential carrier to rear axle housing.

1. Carrier case cap
2. Differential carrier
3. Differential case
4. Differential pinion shaft lock screw
5. Differential pinion gear shaft
6. Thrust washer
7. Differential pinion
8. Main drive ring gear
9. Bearing adjusting ring
10. Thrust washer
11. Side gear
12. Rear axle housing
13. Carrier dowel pin

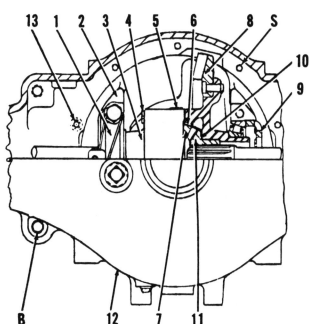

Model RC

245. BEARINGS AND BACKLASH ADJUSTMENT. Carrier bearings are adjusted by turning the adjusting rings (9—Fig. AC111). To adjust the bearings, it is necessary to first detach the rear axle housing (12) from the differential carrier as follows: Support rear of tractor frame and detach rear axle housing from the end of each frame side rail by removing bolts (B) on each side. Remove steering support brace and roll rear axle and torque tube backward as a unit, until it is clear of transmission. Remove both bull pinion shafts from rear axle

housing as outlined in paragraph 277. Remove nuts from studs (S—Fig. AC81 or 111), and pull axle housing away from differential carrier. The carrier bearing adjusting rings (9—Fig. AC111) are now accessible.

245A. Adjust bearings by turning adjusting rings (9). Bearings should be adjusted to a free rolling fit, then tightened one extra notch of one of the adjusting rings to provide a slight bearing preload. Adjusting rings must be set so that differential case is positioned to provide .004-.008 inch backlash between ring and pinion gears. To increase backlash (after bearings have been adjusted), turn adjusting ring on ring gear side "out" and opposite ring "in" the same amount. To decrease backlash, reverse directions. Lock adjustment with ring pin and cotter pin.

246. R & R AND OVERHAUL. To remove the differential assembly, the carrier unit (8—Fig. AC81) or (10—Fig. AC82) must first be removed from the rear axle housing. Support rear of tractor frame and detach rear axle housing from the end of each frame side rail by removing bolts. Remove steering support brace and roll axle and torque tube assembly backward until clear of transmission. Remove bull pinion shafts from the rear axle housing. Refer to paragraph 277 for removal procedure. Remove nuts from studs (S) retaining carrier (2—Fig. AC111) to axle housing (12) and roll axle housing away from carrier and torque tube assembly. Remove carrier

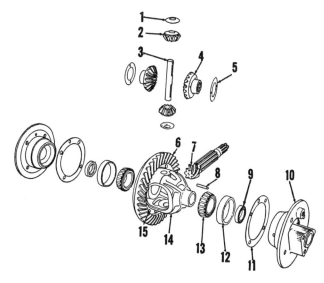

1. Thrust washer
2. Differential pinion
3. Pinion shaft
4. Side gear
5. Thrust washer
6. Bevel ring gear
7. Bevel pinion & shaft
8. Retaining pin
9. Oil seal
10. Bearing carrier
11. Shims
12. Bearing cup
13. Bearing cone
14. Differential case
15. Rivet

Fig. AC110—Model G main drive bevel pinion and ring gear and differential unit. Shims (11) control differential carrier bearing adjustment and bevel ring gear backlash adjustment.

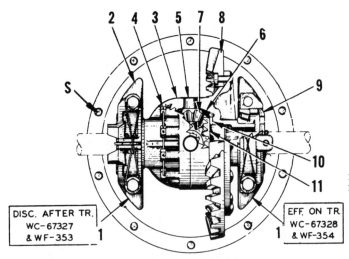

1. Carrier case cap
2. Differential carrier
3. Differential case
4. Differential case cap screw
5. Differential spider
6. Thrust washer
7. Differential pinion
8. Main drive ring gear
9. Bearing adjusting ring
10. Thrust washer
11. Differential side gear

DISC. AFTER TR.
WC-67327
& WF-353

EFF. ON TR.
WC-67328
& WF-354

Fig. AC112—Models WC, and WF bevel ring gear and differential unit assembly—rear view. (S) Nuts & studs retaining differential carrier to rear axle housing.

case caps (1) and remove differential assembly.

246A. To disassemble differential, remove pinion shaft lock screw (4) and pull pinion pin (5) from differential case (3). Pinion gears (7), pinion gear thrust washers, side gears (11), and side gear thrust washers can then be removed from differential case.

246B. Differential bearing cones can be pressed off differential case. Ring gear is retained to differential case with either rivets or heat-treated bolts and nuts. Don't use ordinary bolts. Preferred method of removing old rivets is by drilling them out. When riveting a ring gear to the case, use cold rivets, and be careful to avoid distorting the gear or the case. Run-out of differential unit checked at back face of ring gear should not exceed .003. After unit is reinstalled to carrier, adjust bearings and backlash as outlined in paragraph 245A.

Models WC-WF

247. BEARINGS AND BACKLASH ADJUSTMENT, MODELS WITH MALLEABLE IRON AXLE HOUSINGS. To adjust bearings and gear backlash on these models (on which the final drives are not detachable as complete units from the axle housing), follow the procedure as given for the model RC in paragraphs 245 and 245A. On the WF, it will be necessary to also remove the platform and steering gear.

248. BEARINGS AND BACKLASH ADJUSTMENT, MODELS WITH STEEL AXLE HOUSINGS. To adjust bearings and gear backlash on these models (on which the final drives are detachable at N—Fig. AC122 from the axle housing), first remove both final

drive units. Remove the axle housing from the differential carrier as follows: Support rear of tractor frame and detach rear axle housing from the end of each frame side rail by removing bolts (B). Remove steering support brace on WC, or steering gear and platform on WF, and roll rear axle and torque tube assembly backwards as a unit until it is clear of the transmission. Support the torque tube and remove nuts (S—Fig. AC112) and pull axle housing away from differential carrier. The carrier bearing adjusting ring nuts (9) are now accessible.

248A. Adjust bearings and gear backlash by following the procedure for the RC as given in paragraph 245A.

249. R & R AND OVERHAUL. To remove the differential from its carrier on models with malleable iron axle

housing, follow the procedure given for the model RC in paragraph 246. On models with steel axle housings, follow the procedure given in paragraph 248.

With axle housing moved away from differential carrier remove the carrier bearing caps and lift out the ring gear and differential unit.

To disassemble the differential, remove cap screws (4—Fig. AC112) from case (3). To renew the case or carrier bearings or bevel ring gear, proceed as given for the model RC in paragraph 246B. After differential is reinstalled to carrier, adjust the bearings and backlash by following the procedure given for the RC in paragraph 245A.

Models WD-WD45

250. BEARINGS AND BACKLASH ADJUSTMENT. To adjust bearings and gear backlash, it is necessary to first remove both final drive units as outlined in paragraph 283. Refer to Fig. AC113A. After final drive units are removed from transmission, adjust the bearings and backlash as given for the models B, C, and CA in paragraph 240A.

251. R & R AND OVERHAUL. Main drive bevel ring gear and differential unit are carried in the transmission housing rear compartment. To remove the differential unit, proceed as follows: Drain transmission and differential housing. Remove brake shoes. Remove both final drive units from transmission by removing cap screws at transmission end of units. Unbolt power take-off shaft bearing carrier, and remove power lift arms and shaft assembly. Remove transmission housing rear cover. Support ring gear and dif-

Fig. AC113A—Model WD showing differential unit bearing carrier when left final drive unit is removed. Model WD45 is similar.

Fig. AC113B—Model WD bevel ring gear, and differential unit assembly when viewed through transmission rear cover opening. Model WD45 is similar.

ferential unit and remove differential bearing carriers (1—Fig. AC114). Lift differential unit out through transmission housing rear opening.

252. To disassemble the two pinion, open case type differential unit, proceed as follows: Remove pinion shaft retaining pin (14) and bump pinion shaft (13) out of differential case. Differential pinions (5) and side gears (4) can be renewed at this time.

Recommended backlash of .005 between pinions and side gears is controlled by the copper plated steel side gear thrust washers (3) and/or pinion thrust washers (6). Early production pinions are equipped with bushings, but the bushings are not available for service. Later production pinions do not have bushings. When servicing pinions with bushings, renew pinion shafts and use later type pinions without bushings. New oil seals (8) of treated leather should be installed in bearing carriers with the lip facing inward.

The bevel ring gear, available separately from bevel pinion, is riveted to the one-piece differential case. The preferred method of removing the rivets is to drill them out. When re-riveting, temporarily bolt the ring gear to the differential case and use cold rivets, being careful not to distort the case in the process. Check trueness of ring gear back face by mounting the unit in its normal position in the differential compartment. Total run-out should not exceed .003.

Inner end of each bull pinion gear

shaft is supported by and rotates in a bronze bushing (10) in the differential case. These presized bushings can be renewed without final sizing if carefully installed.

Reinstall bevel ring gear and differential unit and adjust the bevel ring gear to bevel pinion backlash and differential carrier bearings as outlined for models B, C, and CA in paragraph 240A.

MAIN DRIVE BEVEL GEARS

On all models listed in this manual, the mesh position of the main drive bevel is fixed and non-adjustable. The pinion shaft bearings are adjustable for end play on all listed models. On all listed models, the backlash of these bevel gears is controlled by the adjustable differential carrier bearings.

Models B-C

253. **PINION BEARINGS ADJUSTMENT.** To adjust the pinion shaft bearings, first detach torque tube from

1. Bearing carrier
2. Shims
3. Thrust washer
4. Side gear
5. Differential pinion
6. Thrust washer
7. Rivet
8. Oil seal
10. Bushing
11. Bevel ring gear
12. Differential case
13. Pinion shaft
14. Retaining pin
15. Bearing cone
16. Bearing cup

transmission as outlined in paragraphs 185 and 185A. Unstake nut (38—Fig. AC88) and turn same to get free rotation with zero end play; then restake the nut.

253A. **CARRIER BEARINGS ADJUSTMENT.** To adjust carrier bearings and backlash of gears, proceed as outlined in paragraphs 240 and 240A.

253B. **PINION & RING GEAR RENEWAL.** Proceed as outlined in paragraph 203 which covers the renewal of the pinion, and the removal of the ring gear and differential. To renew the ring gear or to overhaul the differential on the bench, follow the procedure in paragraph 241A.

Model CA

254. **PINION BEARINGS ADJUSTMENT.** Adjustment of pinion shaft bearings is controlled by shims (39—Fig. AC90) located on the pinion shaft between the front bearing cone (35) and the 2nd speed gear (32). To add or remove shims it is necessary to remove the bearing cone which usually fits the shaft so tightly that it can be removed only by bumping the pinion shaft rearward and out of the cone. This method of getting to the shims necessitates the removal of the transmission from the tractor, and of the differential from the transmission as described in paragraph 206. The pinion shaft bearings can then be adjusted by following the procedure described in paragraph 211A.

254A. **CARRIER BEARINGS ADJUSTMENT.** To adjust the carrier bearings and the backlash of the bevel gears proceed as described in paragraphs 240 and 240A.

254B. **PINION & RING GEAR RENEWAL.** To renew the pinion, and to remove the ring gear and differential, follow the procedure outlined in paragraphs 209 and 210. The procedure for bench overhaul of the differential, including renewal of the ring gear is described in paragraph 241A.

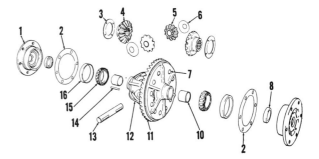

Fig. AC114—Model WD main drive bevel ring gear and differential unit. Shims (2) control differential carrier bearing adjustment and bevel ring gear backlash adjustment.

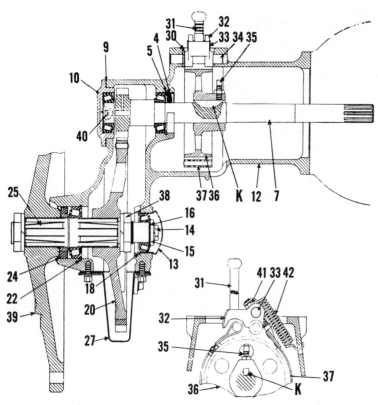

Fig. AC115—Model B final drive and brake assembly. Shims (9) control bull pinion shaft bearings adjustment. Shims (16) control wheel axle shaft bearings adjustment. Model C is similar except for differences in wheel axle shaft construction; refer to Fig. AC116.

K. Brake drum key	16. Shims	33. Brake toggle pin
4. Oil Seal	18. Snap ring	34. Brake fulcrum pin
5. Snap ring	20. Bull (ring) gear	35. Set screw
7. Bull pinion & shaft	22. Snap ring	36. Brake drum
9. Shims	24. Oil seal	37. Brake band
10. Bearing retainer	25. Wheel axle	38. Gear retaining nut
12. Final drive housing	27. Housing pan	39. Rear wheel
13. Bearing cap	30. Brake pin retainer	40. Pinion shaft plug
14. Bearing cap screw	31. Brake lever	41. Adjusting nut
15. Bearing washer	32. Brake toggle	42. Brake band spring

Model G

255. PINION BEARINGS ADJUSTMENT. To adjust the pinion shaft bearings, first separate the main frame from the front of the transmission case as described in paragraph 217. Unstake nut (14—Fig. AC96) and turn same until a slight preload of about 6 inch pounds is obtained; then restake the nut.

255A. CARRIER BEARING ADJUSTMENT. To adjust the carrier bearings and the backlash of the bevel gears, follow the procedure outlined in paragraph 242.

255B. PINION & RING GEAR RENEWAL. To renew the pinion and to remove the differential and ring gear, follow the procedure given in paragraph 222. To overhaul the differential on the bench or to renew the ring gear follow the procedure given in paragraph 241A or 244.

Model RC

256. PINION BEARINGS ADJUSTMENT. To adjust the pinion shaft bearings, it is necessary to first detach rear axle housing from the end of each frame side rail by removing bolts (B—Fig. AC111). Next, remove cap screws (Z—Fig. AC82) retaining torque tube to differential carrier and slide torque tube forward far enough to permit rotation of nuts (12) and (13). Adjust nuts (12) and (13) until pinion shaft rotates freely without end play.

256A. CARRIER BEARINGS ADJUSTMENT. To adjust carrier bearings and backlash of gears, proceed as outlined in paragraphs 245 and 245A.

256B. PINION & RING GEAR RENEWAL. Paragraphs 246, 246A, and 246B give the procedure for installing a new ring gear and overhauling the differential. However in removing the ring gear and differential unit on models with steel rear axle housings, you should deviate from the procedure in paragraph 246 by removing the final drives as complete units instead of removing the bull pinion shafts. After the differential unit is removed a new bevel drive pinion can be installed by detaching the torque tube from the differential carrier by removing nuts (Z—Fig. AC82). Withdraw pinion cage from differential carrier and install the new drive pinion.

Models WC-WF

257. PINION BEARINGS ADJUSTMENT. To adjust the pinion shaft bearings follow the procedure given in paragraph 256.

257A. CARRIER BEARINGS ADJUSTMENT. To adjust carrier bearings and backlash of gears on models with malleable iron rear axle housings, proceed as outlined in paragraphs 245 and 245A. On models having steel rear axle housings, follow the procedure as outlined in paragraph 248.

257B. PINION & RING GEAR RENEWAL. To renew the pinion and bevel ring gear follow the procedure given in paragraph 256B.

Models WD-WD45

258. PINION BEARINGS ADJUSTMENT. To adjust the pinion shaft bearings, follow the procedure outlined in paragraph 234 or 239B. These procedures call for removal of the pinion shaft unless a special wrench is available.

258A. CARRIER BEARINGS ADJUSTMENT. To adjust carrier bearings and backlash of gears, first remove both final drive units as outlined in paragraph 283. Then proceed as outlined for models B, C, and CA in paragraph 240A, to backlash limits of .005-.014.

258B. PINION & RING GEAR RENEWAL. Paragraph 234 or 239B gives the procedure for installing a new bevel pinion, and for removing the ring gear and differential unit from the tractor. Paragraph 252 gives the bench procedure for installing a new ring gear. Paragraph 240A gives the procedure for adjusting the carrier bearings and bevel gear backlash.

FINAL DRIVE UNITS

Models B-C

265. ADJUST WHEEL AXLE SHAFT BEARINGS. Support tractor and remove cap (13—Fig. AC115) from inner side of final drive housing (12). Remove inner bearing retaining cap screw (14), washer (15), and vary the number of shims (16) to eliminate all bearing play but without binding shaft. Removing shims reduces bearing play.

266. RENEW WHEEL AXLE BEARINGS, OR BULL GEAR. Support center of tractor and remove wheel. The oil seal (24) on model B can now be

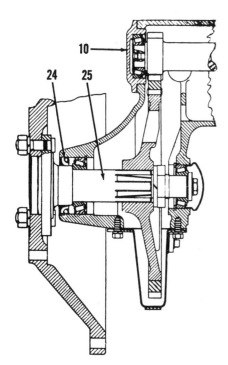

Fig. AC116—Model C details of final drive wheel axle shaft and bull gear. Bearing retainer (10) must have oil grooves when used with a twelve tooth bull pinion.

24. Wheel axle oil seal   25. Wheel axle shaft

removed from the housing. Remove pan (27), cap (13), inner bearing retaining cap screw (14), washer (15), and shims (16). Back off gear retaining nut (38) as far as possible and, using a soft drift, drive axle towards wheel side until gear retaining nut can be unscrewed completely. Withdraw axle shaft and outer bearing cone through wheel side of housing and bull gear (20) nut (38) and washer from below. Bearing cups are positioned in housing by snap rings. Inner bearing cup can be driven out with a driver inserted from the wheel side of the housing and the outer bearing cup with a driver inserted from the inner side.

Reassembly is reverse of disassembly. Select shims (16) between inner bearing retaining washer and axle shaft to remove all bearing play without causing shaft to bind.

267. ADJUST BULL PINION SHAFT BEARINGS. Support tractor and remove wheel. Remove brake housing cover to permit checking bearing play by prying against edge of brake drum (36—Fig. AC115). Remove bearing retainer plate (10) from final drive housing (12). Vary number of shims (9) and paper gaskets to eliminate all bearing play, but without binding shaft. Removing shims reduces bearing play. Alternate shims with paper gaskets to prevent oil leaks.

268. RENEW BULL PINION SHAFT, BEARINGS OR BRAKE DRUM. To remove bull pinion shaft, it is necessary to first remove final drive unit as outlined in paragraph 269.

268A. With final drive unit off the tractor, loosen set screw (35) and, with the aid of an extended puller, pull brake drum (36) from inner end of the shaft and remove brake drum key (K). Remove bearing retainer (10) from final drive housing (12) and withdraw shaft and bearing cones from outer end of housing. Bearing cones can be pressed off shaft. Inner bearing cup is positioned in final drive housing by snap ring (5) and can be driven out after removing oil seal (4). Be sure to renew oil seal in differential carrier bearing housing, if splined end of shaft where it enters seal, shows oil leakage.

Reassembly is reverse of disassembly. Select shims (9) and paper gaskets between bearing retainer and housing to remove all bearing play without causing shaft to bind. Alternate shims with paper gaskets to prevent oil leaks.

269. R & R ONE FINAL DRIVE UNIT. Remove seat assembly and fender. Support center of tractor and remove cap screws retaining final drive housing to transmission and differential housing. Withdraw assembly, being careful not to damage differential carrier oil seal (3—Fig. AC106).

Model CA

270. ADJUST WHEEL AXLE SHAFT BEARINGS. Follow same procedure as for models B and C as outlined in paragraph 265. Refer to Fig. AC119.

271. RENEW WHEEL AXLE, BEARINGS, OR BULL GEAR. Drain gear housing and remove housing pan. Remove rear wheel and tire as a unit. Remove wheel axle shaft bearing dust cap (13), cap screw (14), washer (15), and shims (16). Working from below, remove bull gear positioning snap ring (19). Support bull gear and bump wheel axle shaft out of bull gear and housing.

Oil seal (24) and bearings can be renewed at this time. Lip of oil seal should face towards bull gear. Adjust wheel axle shaft bearings to zero end play, or to a barely perceptible (.002) preload by varying the shims (16) located at inner end of shaft.

272. ADJUST BULL PINION SHAFT BEARINGS. Follow same procedure as for models B and C as outlined in paragraph 267.

Fig. AC117—Model CA tractor—rear view showing the final drive and accessory units installation. Accessory units include the power take-off (67), belt pulley (69), and hydraulic lift pump (66).

12. Final drive housing	63. Hydraulic lift control quadrant	65. Hydraulic ram
47. Final drive clutch actuating lever	64. Hydraulic lift hand control lever	67. Power take-off shaft
62. Lift (rock) shaft		68. PTO housing
		69. Belt pulley shaft

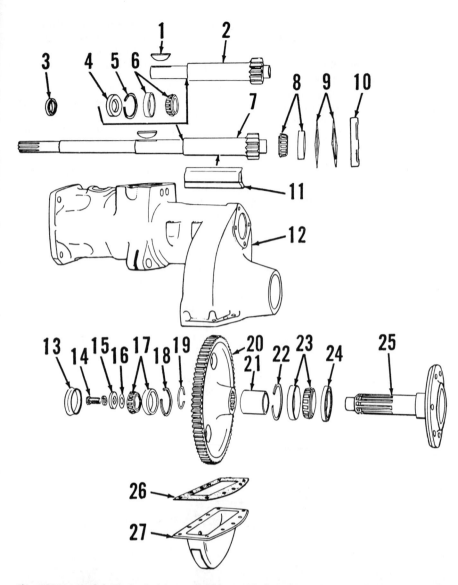

Fig. AC119—Model CA final drive assembly—exploded view. The final drive unit on the right side of PTO equipped models differs from the left side in that it has a two-piece bull pinion (differential) shaft (2) (outer only shown) the pieces of which are connected and disconnected to each other by a Lambert clutch. Right unit bull pinion shaft for models not equipped with PTO is similar to the left bull pinion shaft (7).

1. Woodruff key	9. Shims & gaskets	14. Cap screw	20. Bull gear
3. Oil seal (diff. carrier)	10. Bearing retainer	15. Washer	21. Gear spacer
4. Oil seal	11. Pinion oil tube	16. Shims	22. Snap ring
5. Snap ring	12. Final drive hsg.	18. Snap ring	24. Oil seal
	13. Dust cap	19. Snap ring	25. Wheel axle shaft

273. RENEW BULL PINION SHAFT, BEARINGS, OR BRAKE DRUM. The final drive unit on the right side of pto-equipped models differs from the left side unit in that it has a two-piece bull pinion shaft the halves of which are connected and disconnected to each other by a Lambert clutch. This clutch, which disconnects the engine from the final drive, permits power to be applied constantly to the pto even when the tractor is motionless with respect to the ground.

To remove right side or left side bull pinion shaft, it is necessary to first remove the appropriate final drive as a unit, as outlined in paragraph 274.

With final drive unit off the tractor, the overhaul procedure for the left side unit of a pto-equipped tractor or for either side of tractors not equipped with pto is the same as on models B and C, as outlined in paragraph 268A.

For the right side final drive, the procedure when same is off the tractor is as follows: Remove inner pinion shaft (41—Fig. AC129) from final drive clutch by withdrawing it with the hands. Remove snap ring (53) retaining the clutch unit to the outer shaft (2) and then press or pull the clutch from the shaft. Remove the Woodruff key (1) from the shaft; also the outer bearing retainer (10—Fig. AC119) and

shims (9). The outer shaft can now be bumped out of the housing. The inner bearing cup (6) and housing oil seal (4) can now be renewed. Be sure to renew the oil seal (3) located in the differential carrier bearing housing on the transmission, if inner shaft shows oil leakage. The lip of this seal should face inward; lip of the seal in final drive housing should face outward.

When unit is reassembled, adjust the bull pinion shaft bearings so that shaft rotates freely without any perceptible end play by varying the shims (9) interposed between the final drive housing and pinion shaft bearing retainer.

274. R&R FINAL DRIVE UNIT. To remove one or both final drive units, proceed as follows: Drain housing and support rear portion of tractor under the transmission housing. Disconnect left wheel guard tail light wire at the instrument box. Disconnect pedal shaft by removing clevis pin from clevis on left hand brake rod. Disconnect right hand pedal from its linkage by loosening setscrew in left pedal hub and moving both pedals toward left side of tractor. Disconnect and remove rods from brake bands. Remove tractor seat, seat bar, and shock absorber unit. Remove both wheel guards, rockshaft and supports, and hydraulic lift control quadrant. Remove platform and tire and wheel unit. Disconnect hydraulic ram from final drive housing which is to be removed. Remove nuts retaining final drive housing to transmission housing and withdraw unit from tractor.

Models RC-WC-WF

Some RC and WC tractors are equipped with steel rear axle housings. These tractors are designated by a letter "S" after the serial number and can be identified by examining the outer end of the axle tube. The axle tube of a steel axle housing is attached to the final drive or brake housing by nuts (N) and cap screws, as in Fig. AC122, whereas the axle housing and final drive housings are integral in the malleable iron type.

275. ADJUST WHEEL AXLE BEARINGS. Support tractor and remove hub cap. Turn adjusting nut (16—Fig. AC123) in until all bearing play is eliminated, but without binding the bearings.

276. RENEW WHEEL AXLE, BEARINGS OR BULL GEAR. To remove bull gear (14), support tractor under frame and remove rear wheel. Cork seal (13) can be removed from gear cover at this time by removing retaining nuts. Remove gear cover (11), hub cap, and bearing adjusting nut (16). Slide gear and hub bearings off wheel axle shaft (9). Wheel axle shaft can be driven out of rear axle housing (7) after removing lockscrew

(8). Hub bearing cups can be driven out of drive (bull) gear hub. Seal journal on gear hub is provided with reverse threads to assist in retaining lubricant. Do not interchange left and right gears as threads will then cause leakage of lubricant.

277. RENEW BULL PINION SHAFT, BEARINGS, OR BRAKE DRUM. To remove bull pinion shaft, support tractor under frame and remove bull gear as outlined in paragraph 276. (On shafts having a removable pinion (15—Fig. AC123), same can be pulled from the shaft after removing the retaining nut.) Remove gear cover back plate (10) and brake housing cover (1—Fig. AC138). Loosen brake band adjusting nut (2) and extract shaft by prying outward with a bar placed between brake drum and brake compartment wall. The brake drum can be pressed or driven off the shaft after removing set screw (13). Remove brake drum key (5) and bearing cage (2—Fig. AC123). Bull pinion shaft bearing can be pressed or driven off the shaft after removing snap ring (3).

278. R&R FINAL DRIVE UNIT. On tractors with malleable iron rear axle housing, the final drive is not removable as a unit. On models with steel rear axle housing, remove nuts and cap screws (N—Fig. AC124) attaching the brake housing to the rear axle tube and remove the entire unit from rear axle housing.

Models WD-WD45

279. ADJUST WHEEL AXLE BEARINGS. Adjust bearings to a .002 preload with shims (15—Fig. AC125) interposed between wheel axle shaft retaining cap screw washer (14) and inner end of shaft.

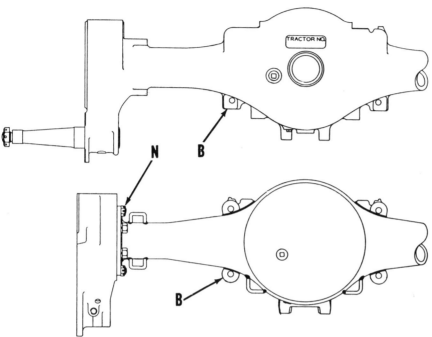

Fig. AC122—Models RC, WC, and WF differential and final drive housing construction. Top view—One-piece malleable iron differential and final drive housing. Bottom view—One-piece stamped steel differential housing to which the final drive housings are bolted on the outer ends. Tractors equipped with steel type housings have a letter "S" following the tractor serial number. Refer to Figs. AC123 & AC124.

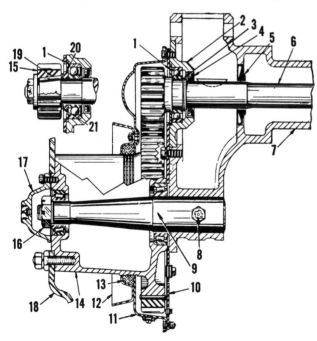

Fig. AC123—Models RC, WC, and WF equipped with a final drive assembly of the malleable iron type. Refer to Fig. AC124 for steel type final drive construction.

1. Bearing retainer	6. Bull pinion & shaft
2. Pinion shaft bearing cage	7. Final drive hsg.
3. Snap ring	8. Axle lock screw
4. Oil seal (outer)	9. Wheel axle shaft
5. Oil seal (inner)	10. Gear cover back plate

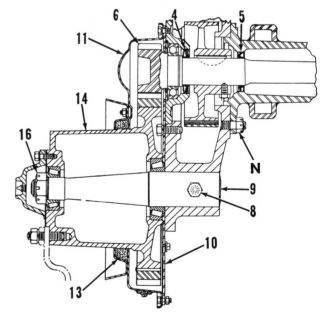

Fig. AC124—Final drive assembly of tractors equipped with a stamped steel type differential housing. "N" is one of the nuts attaching the final drive housing to the differential housing.

11. Gear cover	16. Bull gear retaining nut
12. Hub oil seal retainer	17. Hub cap
13. Seal	19. Pinion gear key
14. Bull gear & wheel hub	18. Wheel
15. Bull pinion	20. Bearing spacer
	21. Spring washer

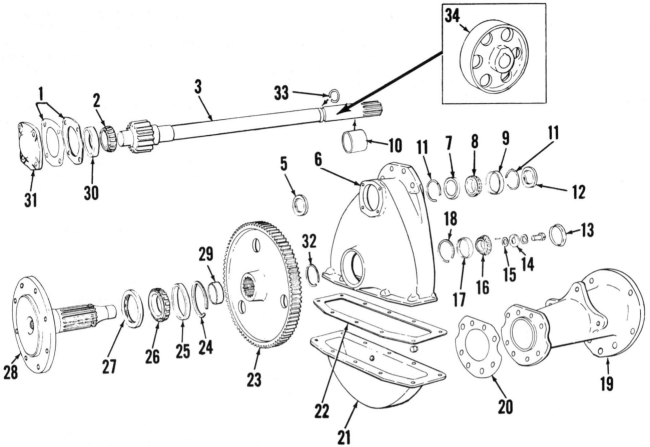

Fig. AC125—Models WD and WD45 final drive and rear axle unit. Shims (1) control bull pinion shaft bearing adjustment. Shims (15) control wheel axle shaft bearing adjustment. Spacer (7) and snap ring (11) are not used on WD tractors after 32831 nor any WD45 tractors.

1. Shims	11. Snap ring, inner bearing	19. Bull pinion shaft housing	27. Oil seal
3. Bull pinion & shaft	12. Oil seal	20. Gasket	28. Wheel axle shaft
5. Oil seal, outer	13. Dust cap	21. Bull gear housing pan	29. Gear spacer
6. Bull gear housing	14. Washer & pin	22. Gasket	31. Bearing retainer
7. Spacer, inner bearing cup	15. Shims	23. Bull gear	32. Snap ring
10. Pinion shaft bushing (in differential case)	16. Bearing cone, inner	24. Bearing cone, outer	33. Brake drum snap ring
	18. Snap ring	26. Bearing cone, outer	34. Brake drum

280. RENEW WHEEL AXLE, BEARINGS, OR BULL GEAR. To renew either the wheel axle shaft (28), bull gear (23) and/or wheel axle shaft oil seal (27) or bearings (16 & 26), proceed as follows: Drain bull gear housing and remove housing pan, and rear wheel and tire unit. Remove wheel axle shaft bearing dust cap, cap screw, washer (14) and shims (15). Working through bull gear housing opening remove bull gear positioning snap ring (32). Support bull gear and bump wheel axle shaft out of bull gear and housing.

The oil seal (27) (lip facing bull gear) and/or bearings can be renewed at this time. Long hub of gear should face toward rear wheel. Adjust wheel axle shaft bearings to a .002 preload with shims (15) located on inner end of shaft.

281. ADJUST BULL PINION BEARINGS. Remove rear wheel and tire as a unit. Adjust bull pinion shaft bearings to provide zero end play and a free rolling fit with shims (1), interposed between bull gear housing (6) and bull pinion bearing retainer (31).

282. RENEW BULL PINION, BEARINGS, BRAKE DRUM. To renew either the bull pinion gear (integral with shaft), pinion bearings, oil seals, brake drum and/or bull gear housing, proceed as follows: Remove brake shoes. Support rear portion of tractor under main frame and remove rear wheel and fender. Supporting the final drive assembly, remove the cap screws which retain it to the transmission and withdraw the final drive assembly.

Brake drum can be renewed at this time by using a suitable puller. Bull pinion shaft housing (19) can be renewed at this time.

Remove brake drum Woodruff key and brake drum positioning snap ring. Remove outer bearing retainer (31) and shims (1). Then bump bull pinion shaft on inner end and remove same from housing.

Adjust bull pinion shaft bearings to provide zero end play and a free-rolling fit by means of shims (1).

283. R&R FINAL DRIVE UNIT. Support rear portion of tractor at main frame and remove rear wheel and wheel guard. Remove brake shoes from same side of tractor. Remove the cap screws which retain final drive unit to transmission case and withdraw the unit from the tractor.

FINAL DRIVE CLUTCH
Model CA

The final drive clutch, Fig. AC126, controls power flow to the rear wheels, but permits the pto and hydraulic pump and belt pulley to operate while tractor is stationary. The single disc, hand operated type clutch is located on the bull pinion shaft of the right hand final drive unit.

284. ADJUSTMENT. Clutch requires adjustment if it fails to disengage the drive to rear wheels when actuating lever is pulled to the rear. To adjust, remove actuator hub retaining screw (40—Fig. AC127) which is accessible through opening in platform and final drive housing. Rotate actuator hub

Fig. AC126 — Model CA tractor Lambert hand operated, self energizing, single plate, dry disc type final drive clutch assembly to provide continuous power take-off, belt pulley, and hydraulic power lift operation. Clutch unit is located on right final drive unit differential (bull pinion) shaft. Right brake drum is integral with clutch.

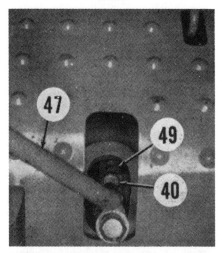

Fig. AC127—Model CA tractor. Final drive clutch can be adjusted through opening in platform and final drive housing by removing adjusting hub cap screw (40), and rotating adjusting hub (49).

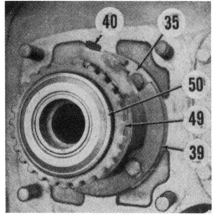

Fig. AC128—Model CA tractor. Final drive clutch actuator support (39), release bearing (50), and adjusting hub (49) is accessible for overhaul after removing the right final drive unit from transmission housing.

35. Differential bearing carrier
40. Adjusting hub cap screw

(49) counter-clockwise (viewed from right side of tractor) until a .005 clearance is obtained between the release bearing and clutch.

285. **R&R AND OVERHAUL.** To remove the assembly, it is necessary to first remove the tractor platform and the right hand final drive unit as outlined in paragraph 274. Release bearing (50—Fig. AC128) can be renewed at this time, using a suitable puller. Remove snap ring (53—Fig. AC129) from shaft; then press or pull clutch from the shaft. To disassemble the unit, remove the nine Allen-head screws (55) which should be unscrewed evenly until spring pressure is relieved. Check the balls and ball seats (56) for wearing or scoring. Renew the clutch hub bronze bushing (54) if running clearance exceeds .012. Renew facings if worn or oil-soaked; also the inner seal (3) in differential carrier if oil leakage is apparent.

285A. When reassembling the unit, use the differential pinion shaft (41) to align the clutch plate splines. Assemble the lined plate (58) with the long hub of same facing the differential unit. To facilitate subsequent adjustment, rotate the actuator hub (49) as far as it will go clockwise (viewed from right side of tractor differential unit.) After final drive is bolted to

Fig. AC130—Model G left rear axle housing installation. Clutch actuating rod passes through left axle housing.

1. Clutch pedal
2. Rear axle housing
3. Pedal return spring
4. Belt pulley cover
5. Clutch fork
6. Clutch rod spring

transmission, readjust the clutch by rotating actuator hub counterclockwise until .005 clearance is obtained between release bearing and clutch. Reinstall the cap screw (40) to lock the adjustment.

AXLE AND HOUSINGS
Model G

286. **AXLE SHAFT AND SLEEVE.** To remove either right or left axle shaft and housing unit, block up tractor, placing the blocking under the transmission case. Remove the rear wheel. Remove the clutch pedal rod for removal of the left axle and hous-

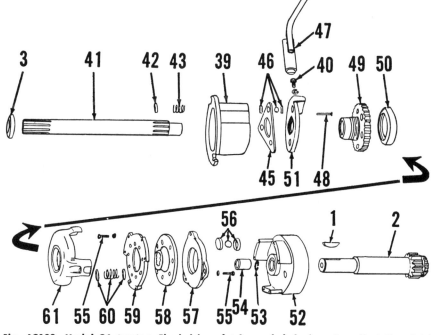

Fig. AC129—Model CA tractor. Final drive clutch—exploded view. To adjust the clutch operating clearance of .005 between release bearing (50) and spring cover (61), rotate adjusting hub (49).

1. Woodruff key
2. Bull pinion shaft (outer)
3. Oil seal (Diff. brg. carrier)
39. Actuator support
40. Adjusting hub cap screw
41. Bull pinion shaft (inner)
42. Spring cup
43. Retracting spring
45. Ball seat
46. Release ball & retainers
47. Operating lever
48. Retaining pin
49. Adjusting hub
50. Release bearing
51. Release actuator
52. Brake drum
53. Snap ring
54. Bushing
55. Allen-head screws (9 used)
56. Disc ball & seats
57. Pressure plate
58. Lined plate (disc)
59. Secondary pressure plate
60. Disc spring & seats
61. Spring cover

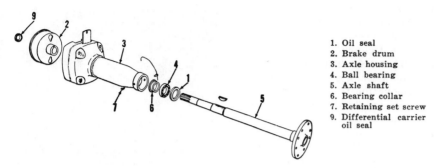

1. Oil seal
2. Brake drum
3. Axle housing
4. Ball bearing
5. Axle shaft
6. Bearing collar
7. Retaining set screw
9. Differential carrier oil seal

Fig. AC131—Exploded view of G rear axle shaft, housing and brake drum unit.

ing. Remove brake pedal return spring (3—Fig. AC130), and bolt from gas tank support. Remove the four nuts retaining axle housing to transmission case and withdraw axle and housing assembly. Reinstall in reverse order.

287. AXLE SHAFT, BRAKE DRUM OR BEARING. To renew shaft (5—Fig. AC131), oil seal (1), or bearing (4), first remove axle shaft and housing unit from tractor. After unit is off tractor, remove the two axle shaft bearing collar retaining screws (7). With a suitable puller or press, remove axle shaft from brake drum (2) and housing (3). Press or drive the bearing collar (6) from axle shaft. At this time, the axle shaft bearing (4) and oil seal (1) may be pressed off of the shaft and renewed. Install treated leather oil seal with the lip facing the bearing. Install new bearing with the shielded side facing the differential unit and lubricate with chassis lubricant. Axle shaft is retained in the housing by the two retaining screws and the press fit of the bearing collar to the axle shaft. For this reason it is advisable to renew the collar when same is removed for either oil seal or bearing renewal.

BRAKES

Models B-C

300. To adjust brakes on later production tractors, turn adjusting screws with a screwdriver inserted through the slot of the brake compartment cover. On tractors prior to Number B52714, it is necessary to first remove the brake compartment cover and then turn the adjusting nut with a wrench.

To remove brake band assembly (37—Fig. AC134), remove fender and brake compartment cover. Remove brake pin retainer (30), spring (42), adjusting nut (41), toggle (32), and toggle pins (33). Pull brake band out, bolt end first. To remove a brake drum, it is necessary to first remove the appropriate final drive unit as described in paragraph 269 after which the drum can be pulled from the bull pinion shaft.

Model CA

301. To take up on brakes, turn adjusting screw (4—Fig. AC135) **in** until a slight drag is obtained when rotating wheel; then back off ½ to ¾ of a turn. Do the same to the other brake.

302. R&R PLATFORM AND BRAKE BAND. To remove either brake band, it is necessary to first remove the tractor platform as follows: Disconnect left wheel guard tail light wire at instrument box. Disconnect pedal shaft by removing clevis pin from clevis on left hand brake rod. Disconnect right hand pedal from its linkage by loosening set screw in hub of left pedal and moving both pedals toward left side of tractor. Disconnect and remove rods from brake bands. Remove tractor seat, seat bar and shock absorber unit. Remove hydraulic lift quadrant and final drive clutch lever. Remove remote ram connection from platform, loosen platform retaining nuts and lift off the platform. Remove wheel guard support.

302A. Raise rear portion of tractor so wheels can be rotated. Remove spring (3) and loosen adjusting screw (4) five or six turns. Remove hairpin locks (1) from the brake band anchor pins then, using a drift punch, carefully push anchor pin out toward wheel. While slowly withdrawing the drift punch, carefully remove the band centering washers (7). If a washer is dropped into the housing, it may be necessary to R&R the final drive unit before it can be retrieved. Remove

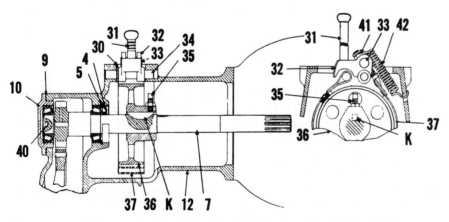

Fig. AC134—Models B, and C brake band and brake drum installation in the tractor final drive unit. Brake drum (36) is located on the bull pinion shaft.

K. Brake drum key
4. Oil seal
5. Snap ring
7. Bull pinion
9. Shims
10. Bearing retainer
12. Final drive housing
30. Brake pin retainer
31. Brake lever
32. Brake toggle
33. Brake toggle pin
34. Brake fulcrum pin
35. Set screw
36. Brake drum
37. Brake band
40. Pinion shaft plug
41. Adjusting nut
42. Brake band spring

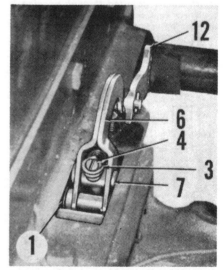

Fig. AC135—Model CA brake installation in left final drive unit. Brakes can be adjusted without removing the tractor platform.

1. Hairpin locks
3. Spring
4. Adjusting screw
6. Toggle
7. Band centering washers
12. Brake lock

the band by pulling up and rearward on toggle end of same.

303. **R&R BRAKE DRUM.** On models equipped with pto, to remove the right hand brake drum (which is integral with the final drive clutch), first remove the right hand side final drive unit as in paragraph 274; then remove the final drive clutch from shaft, and disassemble by removing nine Allen-head screws.

To remove the left hand brake drum, first remove the left final drive unit as outlined in paragraph 274; then remove the drum from the bull pinion shaft using a suitable extended puller.

Use this same procedure for the right hand drum on models with the continuous type pto.

Model G

The 8 inch Bendix brakes, shown in Fig. AC137 are of the internal expanding type, bolted to the differential bearing carrier (10—Fig. AC110). The brake drums are pressed and keyed directly to the wheel axle shafts as shown in Fig. AC131.

304. Each pedal should have approximately 1½ in. of free travel before lining contacts the brake drum. To adjust, raise rear wheel and remove brake rod (2—Fig. AC137) from pedal. To reduce free travel (tighten brake), turn brake rod into brake lever attaching yoke end (1). Both brakes should be adjusted equally.

304A. To remove brake shoes, first remove axle housing and shaft units as outlined in paragraph 286. The shoes can then be detached from their anchorages on the differential carriers. Brake shoes are interchangeable and

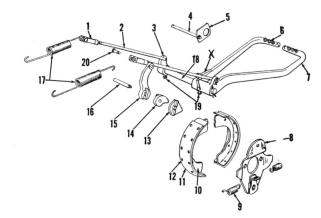

1. Rod adjusting yoke
2. Brake rod
3. Brake rod lever
4. Pedal lock shaft
5. Pedal lock
6. Left brake pedal
7. Right brake pedal
8. Shoe support plate
9. Shoe return spring
10. Shoe
11. Lining
12. Rivet
13. Brake shoe cam
14. Brake lever cam
15. Operating lever
16. Operating lever stud
17. Pedal return spring
18. Pedal shaft
19. Set screws
20. Rod yoke pin

Fig. AC137—Model G linkage and components for the Bendix brake unit. Pedal stop is shown at point (X).

the top of the shoe may be identified by the cutout section in the cam surface. Install new linings on shoes so that lining ends are flush with lower end of shoe.

304B. To remove brake drum, first remove axle housing and shaft unit, as in paragraph 286; then pull drum from inner end of axle shaft. When new pedals are installed, the pedal stop shown at point "X" must be filed to position the pedals approximately 4 in. below the seat attaching pad and to have the pedals level with each other.

Models RC-WC-WF

305. To take up on brakes, remove fenders and brake lever covers and turn adjusting nut (2—Fig. AC138) until a slight drag is obtained when rotating wheel; then back off ½ to ¾ of a turn. Do the same to the other brake.

305A. To remove brake band assembly, remove fender and brake lever

cover. Remove the adjusting nut (2) also the spring (9). Push bolt out of toggle (3). Remove cotter pin from brake lever shaft (10) and bump shaft out of the housing key against inner bushing (12). Continue bumping until shaft and bushing are out of housing. Brake band can now be withdrawn from housing. It may be necessary to install a new inner bushing (12) due to possible damage during its removal.

305B. To remove brake drum, follow the procedure outlined in paragraph 277.

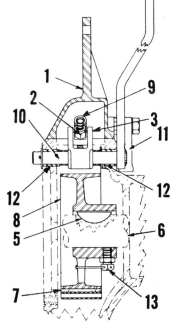

Fig. AC138—Models RC, WC, and WF brake unit which is located on each final drive bull pinion shaft.

1. Housing cover	8. Drum
2. Adjusting nut	9. Band spring
3. Band toggle	10. Lever shaft
5. Drum key	11. Lever lock
6. Bull pinion shaft	12. Lever shaft bushing
7. Brake band	13. Drum set screw

1. Hairpin locks
2. Band pivot pin
3. Spring
4. Adjusting screw
5. Toggle pin
6. Toggle
7. Band centering washers
8. Brake drum key
9. Brake drum
10. Brake band
11. Band anchor pin

Fig. AC136—Model CA brake assembly—exploded view. Brake drum (9) is used in both right and left units for tractors not equipped with a continuous type pto. For models equipped with a continuous type pto, the right unit brake drum is integral with the final drive clutch.

Fig. AC139—Model CA left unit brake installation when platform and wheel guard and support are removed. Brake band can be removed at this time.

1. Hairpin locks
2. Band pivot pin
3. Spring
4. Adjusting screw
5. Toggle pin
6. Toggle
7. Band centering washer

Models WD-WD45

The brakes are of the external contracting type, Fig. AC140, with each brake drum keyed to the inner end of its bull pinion gear shaft.

306. To adjust either brake unit turn screw (18) in until slight drag is obtained when turning wheel; then back off screw ½ turn. Turn nut on brake shoe clevis (6) until slight wheel drag is obtained, then back off about ½

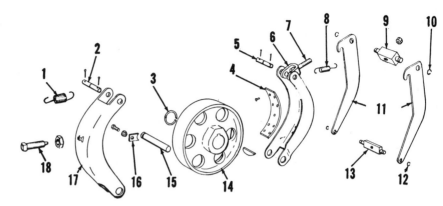

Fig. AC140—Model WD brake unit. Brake drums are keyed to inner end of each final drive bull pinion shaft. On late WD tractors, pin (15) is provided with a head. Model WD45 is similar.

1. Shoe return spring (rear)	7. Adjusting bolt	13. Toggle pin
2. Pin	8. Shoe return spring (front)	14. Brake drum
3. Drum snap ring	9. Adjusting toggle pin	15. Shoe anchor pin
4. Brake lining	10. Snap ring	16. Anchor pin lock
5. Pin	11. Toggles	17. Brake shoe
6. Clevis	12. Snap ring	18. Shoe centralizing screw

turn. Do the same to the other brake.

Each brake pull rod should measure 16 5/16 inches from center line of hole in brake pedal to front face of toggle (13). This setting provides best pedal leverage.

306A. To remove brake shoes, remove brake compartment covers. Disconnect and remove front and rear shoe return springs (1 & 8). From lower side of transmission housing, remove brake shoe anchor pin lock, and remove anchor pin (15). In the event that the brake anchor pin cannot be removed by prying, it may be necessary to drill and tap the anchor pin so that a suitable puller of the stud

and sleeve type may be used. Remove adjusting nut from clevis (6). Front shoe should be removed first, through compartment cover opening. Remove rear shoe in like manner.

Reinstall the front shoe, as an assembly made up of the brake shoe, toggle levers, toggle pin and adjusting clevis and pin. The large brake shoe return spring (1) is for the rear shoe and the smaller spring (8) is used for the front shoe.

306B. To remove a brake drum, it is necessary to first remove its final drive unit and pull drum off bull pinion shaft as outlined in paragraph 282.

BELT PULLEY and PTO

BELT PULLEY UNIT

G Without Hydraulic Lift

This model is furnished with a belt pulley unit only, or with a combined belt pulley and hydraulic lift unit. It is not furnished with a conventional power take-off. Refer to paragraphs 309A and 309B for repair data on models equipped with hydraulic lift.

308. **COMPLETE ADJUSTMENT.** Both shafts of the unit are carried on tapered roller bearings which should be in correct adjustment **before** attempting to adjust the backlash of the belt pulley gears. Bearings of both shafts are considered in correct adjustment when the shafts rotate freely but have zero end play.

308A. To adjust bearings of bevel pinion shaft (10—Fig. AC145), turn the nut (2) located under dust cap at pulley end of shaft. After completing

the adjustment, lock the nut (2) by staking thin portion of same into the shaft keyway.

Adjustment of bearings for the bevel gear shaft (13) is accomplished by re-

moving or adding shims (7) located under the two bearing retainers.

Belt pulley gears are adjustable for backlash, but the mesh position is fixed and non-adjustable. Backlash

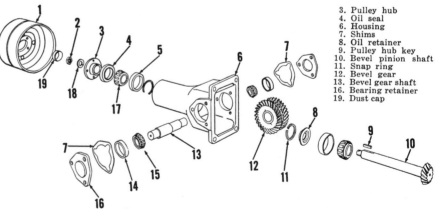

3. Pulley hub
4. Oil seal
6. Housing
7. Shims
8. Oil retainer
9. Pulley hub key
10. Bevel pinion shaft
11. Snap ring
12. Bevel gear
13. Bevel gear shaft
16. Bearing retainer
19. Dust cap

Fig. AC145—Model G belt pulley unit which is supplied for tractors not equipped with a hydraulic lift system. Recommended backlash of .005-.007 between bevel gear (12) and pinion (10) is controlled with shims (7) which also control bearings adjustment.

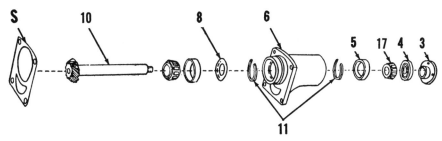

Fig. AC146—Model G belt pulley unit which is supplied for tractors equipped with a hydraulic lift system.

S. Shims	5. Bearing cup	10. Bevel pinion shaft
3. Pulley flange	6. Pulley sleeve	11. Snap ring
4. Oil seal	8. Oil retainer	17. Bearing cone

between bevel pinion (10) and gear (12), should be adjusted only **after** it is known that the bearings are correctly adjusted, as outlined above. Recommended backlash is .005 to .007, which is obtained by removing a shim or shims (7) from under one bearing cap and installing the same under the opposite bearing cap.

309. R & R AND OVERHAUL. To remove belt pulley and housing unit, unscrew the four cap screws which retain it to the clutch housing.

To disassemble the unit remove dust cap from pinion shaft and the bearing adjusting nut (2). Remove bevel gear shaft bearing caps (16), move shaft (13) away from pinion (10) as far as possible, then insert close fitting blocks to fill the space between bevel gear and case on the pinion side of gear. Press shaft (13) from gear and remove from case. The pinion shaft can now be pulled or bumped from the case.

When reassembling, install the pinion shaft first with lip of oil seal for same pointing towards the pinion. Adjust pinion shaft bearings as in paragraph 308A. When assembling the bevel gear shaft, press shaft in until bevel gear is bottomed on shaft shoulder. Adjust bevel gear shaft bearings and backlash as in paragraph 308A.

The paper gasket interposed between the pulley unit housing and clutch housing is .010 thick. If this standard gasket does not provide .004 to .007 backlash between helical drive and driven gears, use a thinner gasket or an additional gasket as required.

The driving gear which meshes with the helical gear shown just behind the bevel gear (12) is located in the clutch housing and can be removed as outlined in paragraph 215.

Model G With Hydraulic Lift

309A. PULLEY AND SHAFT UNIT. This unit, Fig. AC146, can be removed from the larger housing by unscrewing four cap screws. The gear is integral with the shaft (10) and is supported on tapered roller bearings which should be adjusted to free rotation, zero end play, by tightening the nut located under the dust cap at the end of the pulley flange (3). This adjustment can be made with unit in place, but preferably when it is off the tractor. Lip of oil seal (4) should face toward the bevel gear. After reinstalling unit to larger housing, check gear backlash which should be within limits of .005-.010. Obtain desired backlash by varying the shims (S) located between flange face of unit and face of larger housing.

1. Bearing cover	31. Detent ball	41. Valve lever
2. Gasket	32. Needle bearing	42. Valve cover
3. Shaft nut	33. Pump gear (idler) shaft	43. Spring
4. Ball bearing	34. Pump gear (drive)	44. Adjusting screw
5. Shifter	35. Dowel pin	45. Oil seal
6. Shifter pins	36. Spring	46. Control valve
7. Drive shaft	37. Plug	47. Valve cover stud
8. Roller bearing	38. Gasket	48. Ball seat
9. Housing	39. Retainer	49. Ball
10. Shift arm	40. Detent spring	51. Detent ball
11. Shift lug		
12. Snap ring		
13. Belt pulley bevel drive gear		
14. Seal ring		
15. Hydraulic pump intake tube		
16. Gasket		
17. Bushing		
18. Thrust washer		
19. Needle bearing		
20. Pump body		
21. Pump center plate		
22. Drive gear		
23. Gasket		
25. Pump gear (idler)		
26. Seal		
27. Shifter lever (to place BP in operation)		
28. Detent spring		
29. Gasket		
30. Detent retainer		

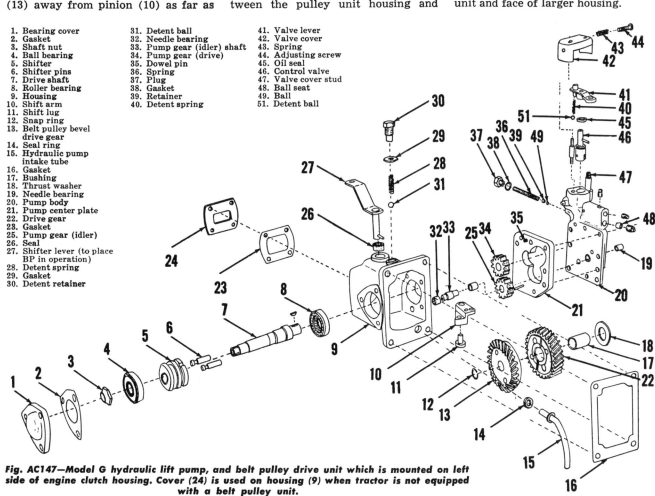

Fig. AC147—Model G hydraulic lift pump, and belt pulley drive unit which is mounted on left side of engine clutch housing. Cover (24) is used on housing (9) when tractor is not equipped with a belt pulley unit.

309B. BEVEL DRIVE GEAR AND HOUSING. To renew the bevel gear, Fig. AC147, which drives the pulley shaft gear, or to overhaul any part of the drive shaft, it is necessary to re- move the drive (larger) housing from the tractor clutch housing. Removal is accomplished by disconnecting the hy- draulic unit hose lines and external control linkage, and unscrewing the cap screws which retain the housing to the clutch housing.

OVERHAUL. Procedure for disas- sembly of the removed unit is as fol- lows: Remove pulley and shaft unit, and pump body (20) retaining cap screws. Remove pump gears (25 & 34) and gear key from drive shaft (7). Re- move shifter lever detent spring (28) and ball (31). Remove drive shaft bearing cap (1) and shaft nut (3). Working through housing opening, bump ball bearing (4) out of housing and off shaft. Remove shifter collar (5) and pins (6) out through bearing opening. Move shaft and gears toward bearing opening until opposite end is freed from roller bearing (8). Tip shaft slightly and withdraw through mounting face opening. Roller bear- ing (8) can now be removed from housing.

Helical drive gear (22) rotates on a presized bronze bushing (17), which should be renewed when the running clearance exceeds .006.

When reinstalling the drive shaft housing unit to the tractor clutch housing, check backlash of drive heli- cal gear with its driving gear in clutch housing. If backlash is not within lim- its of .004-.007, vary the thickness of the gasket (16) to obtain this desired backlash.

To overhaul the hydraulic pump at this time, refer to paragraph 328 for procedure.

Models RC-WC-WD-WD45-WF

The data in paragraphs 310 and 310A apply to only the Belt Pulley. Refer to paragraphs 314 through 314B for repair data applying to the Power Takeoff used on these models.

310. GEAR ADJUSTMENT. A series of radial holes (13—Fig. AC148) pro- vides a means of obtaining the recom- mended .005-.011 (⅛ inch movement of pulley rim) bevel drive gear back- lash. Rotate belt pulley tube counter- clockwise to reduce backlash.

310A. OVERHAUL. To remove belt pulley unit, loosen set screw (located on clutch housing) and withdraw unit.

To disassemble belt pulley unit, re- move pulley retaining nut (8) and pulley. Remove snap ring (1) from gear end of belt pulley tube, and bump shaft out of tube.

Install new oil seal (7) of treated leather with the lip facing the gear. The inner ball bearing (15) which takes thrust should be installed in one of the following ways: Install bearings which are stamped "THRUST" on inner race so that the mark faces the gear; install bearings

Fig. AC148—Model WD belt pulley assembly. Models RC, WC, WD45 & WF are similar.

1. Snap ring
2. Pulley carrier tube
3. Set screw
4. Main frame
5. Belt pulley
6. Ball bearing (outer)
7. Oil seal
8. Pulley retaining nut
9. Snap ring
10. Tube cover
11. Tube seal
12. Belt pulley shaft
14. Snap ring
15. Ball bearing (inner)

Fig. AC149A—Model CA accessory units (belt pulley, power take-off, and hydraulic lift pump) housing, which is attached to the rear face of the transmission housing.

BG. Main drive bevel ring gear & differential unit
MS. Transmission main shaft
TH. Transmission housing
13. Accessory units drive gear shift lever
14. Hydraulic lift pump
15. Accessory units drive gear
16. Snap ring
17. Roller bearing
18. Snap ring
19. Shims
20. PTO shaft
21. Belt pulley shaft
22. Shims
23. Accessory units driven & belt pulley drive gear
24. Shims
25. Differential unit & bull pinion shaft
30. Bearing retainer
31. Snap ring
32. Belville washer
33. Oil seal
34. Oil slinger
35. Snap ring
36. Pulley shaft gear

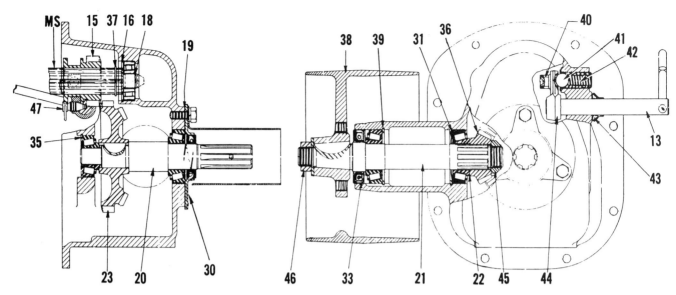

Fig. AC149B—Models B, and C accessory units (belt pulley, power take-off, and hydraulic lift pump) housing, which is attached to the rear face of the transmission housing.

MS. Transmission mainshaft
13. Accessory units drive gear shift lever
15. Accessory units drive gear
16. Snap ring
18. Snap ring
19. Shims

20. PTO shaft
21. Belt pulley shaft
22. Shims
23. Accessory units driven & belt pulley drive gear
30. Bearing retainer
31. Snap ring

33. Oil seal
35. Snap ring
36. Pulley shaft bevel gear
37. Snap ring
38. Belt pulley
39. Snap ring
40. Washer

41. Detent ball
42. Detent spring
43. Seal
44. Shift finger
45. Adjusting nut
46. Pulley shaft nut
47. Shift finger screw

which are stamped "THRUST" on **outer** race so that the mark is facing away from gear.

The one-piece belt pulley tube cork seal (11), which is located between clutch housing and inside of right main frame rail and used to prevent lubricant from leaking past the belt pulley tube, can be renewed in the following manner: Loosen right frame rail bolts and remove oil seal. Cut the new cork seal diagonally and shellac same into position with diagonal joint placed on top.

To renew the pulley unit drive gear (3—Fig. AC84) requires removal of clutch shaft as outlined in paragraph 192.

COMBINED PULLEY & PTO

Models B-C-CA

On these models, the pto is combined not only with the belt pulley, but also with the pump of the hydraulic system. These three accessory units are all contained within or attached to a common housing, Fig. AC149A, designated by Allis-Chalmers as the pto housing. There are detailed differences in the construction of the CA as compared to models B and C, Fig. AC149B, including the use of four cams on the CA for the hydraulic pump instead of two, and a separate set of shims on the CA for controlling the position of the bevel gear on the pto shaft. This separate set of shims is the only item affecting the repair procedures and can be overlooked generally, except in those cases where the pto bevel

shaft gear or the pto housing is being renewed.

Except that the bearings of the pto shaft can be adjusted and the belt pulley shaft oil seal renewed with the pto housing in place, all repair work on either the pto shaft or the belt pulley shaft necessitates the removal of the housing from the rear face of the transmission.

311. **R & R PTO HOUSING.** To remove the pto housing, Fig. AC150, from rear of transmission, drain the housing or jack up rear end of tractor. Loosen drawbar rockshaft and disconnect hydraulic valve control unit from pump. Remove belt pulley. Disconnect pump to drawbar linkage. Remove control quadrant from platform; also tractor seat spring and shock absorber

unit. Loosen drawbar hanger bolts from rear lower face of transmission to provide removal clearance for housing. Remove cap screws and nuts which retain housing to transmission and lift housing unit off tractor.

When reinstalling the housing make sure that the housing to transmission case gasket is partially blanked off to form a dam as shown at (G—Fig. AC151.)

312. **OVERHAUL.** To disassemble the removed unit, first remove the hydraulic pump unit from the case. Remove the bearing retainer and shims from rear face of housing. Using a puller with jaws of same engaged against rear face of spur gear on pto shaft as shown in Fig. AC153 press the pto shaft out of the gear. The front

Fig. AC150—Model CA accessory units installation—right rear view.

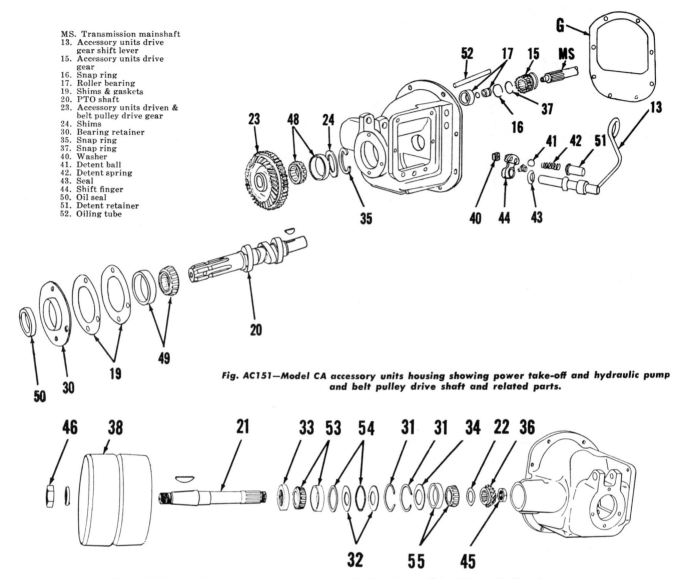

MS. Transmission mainshaft
13. Accessory units drive gear shift lever
15. Accessory units drive gear
16. Snap ring
17. Roller bearing
19. Shims & gaskets
20. PTO shaft
23. Accessory units driven & belt pulley drive gear
24. Shims
30. Bearing retainer
35. Snap ring
37. Snap ring
40. Washer
41. Detent ball
42. Detent spring
43. Seal
44. Shift finger
50. Oil seal
51. Detent retainer
52. Oiling tube

Fig. AC151—Model CA accessory units housing showing power take-off and hydraulic pump and belt pulley drive shaft and related parts.

Fig. AC152—Model CA accessory units housing showing belt pulley shaft and related parts.

21. Belt pulley shaft	32. Belville washer	36. Pulley shaft gear	53. Bearing (outer)
22. Shims	33. Oil seal	45. Adjusting nut	54. Spacer
31. Snap ring	34. Oil slinger	46. Retaining nut	55. Bearing (inner)

bearing cup shims and snap ring will remain in the housing. If this cup is removed for any reason, be sure to retain and mark the shim or shims located between the cup and the snap ring. Remove nut (45—Fig. AC152) and shims from inner end of belt pulley shaft. Bump belt pulley shaft out through side of housing and with it, the oil seal (33) outer bearing cone and shims (22). Bearing cups can now be removed from the housing.

313. During reassembly, observe the following points:

Belville spring washers (32) should be assembled to belt pulley shaft with cupped face of outer one facing outward and cupped face of inner washer facing inward.

Belt pulley shaft bearings should be adjusted by means of nut (45) to a preload of 7-10 inch pounds to rotate

shaft when pto shaft is out of housing. Lip of oil seal should face inward.

If pto shaft front bearing cup was removed, be sure to reinstall the same shims (24—Fig. AC151) as were removed from between it and the snap ring.

After pto shaft bearings have been adjusted to free rotation with zero end play, observe mesh position of pto bevel gear in relation to bevel gear on belt pulley shaft. The heel faces of both should be flush with each other within .006 when the backlash of mating teeth is within the limits of .004-.007. If backlash is less than .004, remove a shim (24) from the belt pulley shaft or add a shim if backlash is greater than .007. If heel faces are not within .006 of being flush with each other after correct backlash has been obtained, it will then be necessary to

change the position to the pto gear by removing or adding shims (24). If this must be done do so by removing the front bearing snap ring (35) to avoid the longer job of disassembling the pto shaft.

POWER TAKE-OFF UNIT
Models RC-WC-WD-WD45-WF

This data applies only to the Power Take-off unit. Refer to paragraphs 310 and 310A for repair data applying to Belt Pulley used on these models.

There are detailed differences in the construction of the pto used on models RC, WC, and WF, as compared to models WD, and WD45, and between the pto used on model WD prior to tractor serial WD-2400 and later WD tractors. The accompanying illustrations, Figs. AC154 & AC155, show any necessary variations in repair procedure.

Fig. AC153—Puller arrangement used in removing the pto shaft from the accessory units housing on a model CA tractor. Models B, and C pto shaft can be removed in a similar manner.

Fig. AC156—Model CA pto shaft when viewed from hydraulic pump mounting pad surface of accessory housing. Note cams on pto shaft which are used to operate the hydraulic pump.

314. **R & R AND OVERHAUL.** The power take-off extension shaft and bearing carrier can be removed for overhaul after unbolting the carrier and removing the coupling bolt or splitting the universal joint at the front end of the extension shaft. The extension shaft front and rear oil seals should be installed with the lips facing the bearing.

314A. The power take-off idler and sliding gear housing can be removed for overhaul after removing the extension shaft and bearing carrier.

Disassembly of the unit is self-evident after an examination and reference to Figs. AC154 & AC155. Steel shims (15), alternated with paper gaskets, (inserted between bearing cage and housing) control the adjustment of the stub shaft bearings. Adjust bearings to provide zero end play and yet permit shaft to rotate freely. Install stub shaft oil seal with the lip facing the gear.

Vary the thickness of gasket or gaskets (30) at sliding gear housing and transmission case to provide .003-.007 of backlash between gear teeth.

314B. For removal and overhaul of the WD and WD45 power take-off drive gear (5—Fig. AC84) which is splined to rear end of engine clutch shaft or of intermediate drive gear (12—Fig. AC86) which are located in torque tube, refer to paragraph 193.

For removal and overhaul of the RC, WC, and WF pto drive gear, which is located on the transmission countershaft, refer to paragraph 228.

POWER LIFT SYSTEM

Most of the troubles encountered with the hydraulic system on modern tractors are caused by dirt or gum deposits. The dirt may enter from the

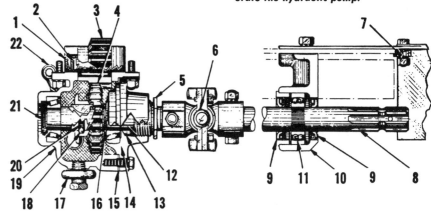

Fig. AC154—Models RC, WC, and WF power take-off unit which is mounted on the lower side of the transmission housing. Idler gear (3) meshes with a gear on the transmission countershaft gear cluster.

1. Idler gear pin	7. Bevel washer	13. Shaft packing	18. Detent ball
2. Gear shaft pin	8. PTO shaft	14. Bearing cage	19. PTO housing
3. Sliding gear	9. PTO shaft oil seal	15. Shim	20. Detent ball spring
4. Packing nut	10. Rear bearing housing	16. Snap ring	21. Expansion plug
5.	12. Stub shaft	17. Housing brace	22. Shift lever
6. Universal joint			

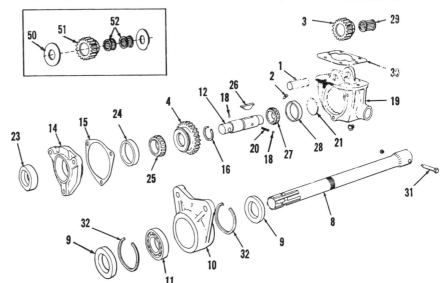

Fig. AC155—Models WD and WD45 power take-off unit which is mounted on the lower side of the torque tube. Items (50, 51 & 52) which supersede items (3 and 29) are effective on WD tractors after serial WD2400 and all WD45 tractors.

1. Idler gear shaft	10. Bearing carrier	18. Detent balls	30. Gasket
2. Pin	11. Bearing	19. PTO drive housing	31. Extension shaft bolt
3. Idler gear	12. Stub shaft	20. Spring	32. Snap ring
4. Sliding gear	14. Bearing cage	23. Oil seal	50. Thrust washer
8. Extension shaft	15. Shims	26. Woodruff key	51. Idler gear
9. Oil seal	16. Snap ring	29. Roller bearing	52. Roller bearing

outside, it may have been in the system originally, or it may show up as the result of wear or partial failure of some part of the system. The presence of gummy deposits however usually results from inadequate fluids or from failure to drain and renew the fluid at the recommended intervals. These principles should be kept in mind when shooting trouble on any of the existing systems and also when performing repair work on pumps, valves and cylinders. Thus when disassembling a pump or valve unit, it is good practice generally to not remove any parts which can be thoroughly inspected while they are installed. Internal parts of pumps, valves and cylinders when removed should be handled with the same care as would be accorded the parts of a diesel pump or injector and should be soaked or manually cleaned with an approved solvent to remove gum deposits. Unless you p r a c t i c e good housekeeping (cleanliness) in your shop do not undertake the repair of hydraulic equipment.

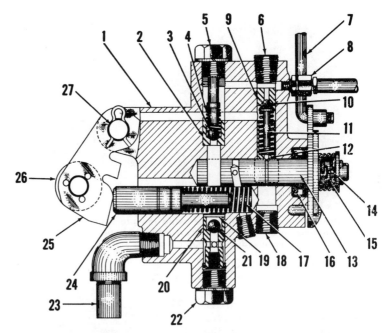

Fig. AC160—Models B, and C power lift hydraulic pump which is of the two plunger type. Pump is mounted on right side of accessory units housing.

1. Pump body	10. Relief valve ball	19. Intake valve insert
2. Discharge valve insert	11. Relief valve spring	pin
3. Discharge valve ball	plunger	20. Intake valve sleeve
4. Discharge valve	12. Relief valve thimble	21. Intake valve ball
spring	13. Control shaft	22. Intake valve plug
5. Discharge valve plug	14. Torsion spring	23. Intake pipe
6. Relief valve plug	15. Retaining pin	24. Plunger
7. Control rod	16. Shaft oil seal	25. Cam arm
8. Outlet connection	17. Plunger spring	26. Cam roller
9. Relief valve insert	18. Access plug	27. Cam arm pin

HYDRAULIC PUMP

MODELS B-C

In cases of faulty operation of the pump valves the unit can oftentimes be corrected without disassembly by removing it from the tractor and flushing it with a petroleum solvent or gasoline.

320. R & R AND INSPECTION. Removal of the pump from the pto housing is accomplished by disconnecting the hose lines and controls and removing the bolts or screws which retain it to the pto housing. Removal of the pump driving camshaft which is integral with the pto shaft is done by following the procedure outlined in paragraphs 311 and 312.

Procedure for bench disassembly of the pump is as follows:

Plungers can be lifted out of their cylinders after removing the cam arm pins (27—Fig. AC160 or AC161) which are retained by hair pin locks.

Inlet valves are removed by unscrewing the valve plugs (22) and pulling the internally threaded seat inserts (37) down with a sleeve and puller screw threaded into the insert. If seat inserts are not threaded they can be driven out using a drift 1/32 inch smaller in diameter than the O.D. of the insert. This method usually calls for a renewal of the insert because of damage done during removal.

Discharge valves (3) can be removed by same methods used on inlet valves after removing valve plugs (5)

springs and balls. These valves are removed through top of housing.

Control shaft (13) is removed by first removing the washers and torsion spring (14). Turn shaft to "lift" position where thimble (12) is at highest point of its travel then withdraw shaft from pump body.

Relief valve insert (9) can be bumped out of top of pump after removing top and bottom plugs, ball, spring and plunger (11).

Plungers and cylinder bores must be free of roughness or deep scratches. Recommended plunger to cylinder clearance is .0002. Check cam followers for wear or roughness and reject any rollers which have more than .010 clearance on journal pin. Check the valve seats and balls for grooves, pitting or other leak producing conditions oil leakage occurring between control shaft and its bore in body will show up at drilled passage between plungers in line with control shaft bore.

321. OVERHAUL AND T E S T. If valve seat inserts were removed by bumping, new inserts will probably be needed.

CONTROL SHAFT. Assemble the control shaft to the lever so that stop for adjusting screw in lever is toward flat on control shaft. Insert shaft into body with flat on shaft up

and with torsion spring wound ½ turn so as to return the control shaft to "hold" position.

RELIEF VALVE. With control shaft in lowering position install in order thimble (12), spring (32) and plunger (11). Before installing the seat insert, seat the ball to same by tapping lightly with a hammer and soft drift. Install the seat insert, drop ball into cupped end of plunger then carefully press seat into body until top of insert (9) is 9/16 inch below top of hole.

DISCHARGE VALVES. Press or bump discharge valve seat insert into pump body until it bottoms. Discharge valve plugs of three different lengths, Fig. AC162, have been used in the three different pump bodies used at various times during the production of the B and C tractors. When the correct plug is used the ball end of same will contact the spring and hold the ball on its seat when plug is screwed all the way down. If ball rattles when pump is shaken or if plug cannot be screwed all the way to its seat on pump body the plug is incorrect for that particular pump. The different plugs are procurable under number 213471 short; 213005 medium; 211360 long.

INLET VALVES. Install seat inserts up into body until top of insert pin just clears the plunger spring (17).

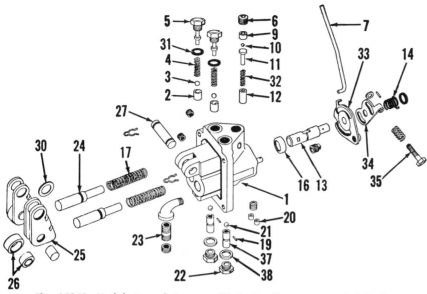

Fig. AC161—Models B, and C power lift hydraulic pump—exploded view.

1. Pump body
2. Discharge valve insert
3. Discharge valve ball
4. Discharge valve spring
5. Discharge valve plug
6. Relief valve plug
7. Control rod
9. Relief valve insert
10. Relief valve ball
11. Relief valve spring plunger
12. Relief valve thimble
13. Control shaft
14. Torsion spring
16. Shaft oil seal
17. Plunger spring
19. Intake valve insert pin
20. Intake valve sleeve
21. Intake valve ball
22. Intake valve plug
23. Intake pipe
24. Plunger
25. Cam arm
26. Cam roller
27. Cam arm pin
30. Washer
31. Gasket
32. Relief valve spring
33. Control plate
34. Friction plate
35. Lowering rate adjusting screw
37. Intake valve insert
38. Gasket

Fig. AC162 — Models B, and C hydraulic pump discharge valves which were used in different style pumps. (A) Part No. 213471 as used in pumps with serial number A1 and up. (B) Part No. 213005 as used in pumps with serial number 2585 to 10114. (C) Part no. 211360 as used in pumps with serial number prior to 2585.

If intake valve plugs will not screw all the way in when insert is clear of plunger spring, install extra copper ring gasket between head of plug and pump body. In some cases the valve plugs can be used as screw pushers to move the seat inserts into position after the inserts have been pressed in far enough to permit engaging the first three threads of the plug.

Seat new or old valves **after** installation (except relief valve) by tapping balls against their seats with a hammer and soft punch.

322. TEST OF PUMP AND VALVES. In the absence of a test fixture for operating the plungers the unit can be tested by reinstalling it to the pto housing. Before doing so however, first turn the pump upside down and prime it by pouring approved fluid into inlet while working plungers by hand. If only one copper washer is used under the pump attaching screws it should be under the head of the long screw in front lower hole in pump. After running pump for a few minutes bleed the system at ram packing nuts with control in raised position. Retighten the packing nuts **before** moving the control to "lowering" position.

Attach pressure gauge to pump outlet. Operate pump with control valve in "lift" position until pressure reaches 3,200 pounds, and relief valve opens. Relief valve opening will be accompanied by a squealing or chattering noise in the pump. If relief valve opens below 3,200 pounds, push relief valve seat deeper into pump so as to increase the effective spring pressure. With pressure at about 3,100 pounds and control valve in "lift" position, stop pump and watch pressure drop. Pressure drop should be slow and should not drop below 1,500 pounds. The same results should be obtained when test is repeated in "hold" position.

With adjusting screw backed out, move control valve to "lowering" position. If pressure does not immediately drop to 100 pounds or less, relief valve seat has been driven too deeply into pump body. Turning adjusting screw in, retards lowering rate.

Models CA-WD-WD45

Pump on model CA is mounted on the side of the main pto housing which is bolted to the rear face of the transmission housing as shown in Fig. AC150. The CA pump is driven by 4 cams which are integral with the pto shaft. On WD and WD45, pump is mounted on the right side of the torque tube, as shown in Fig. AC163 and receives its drive from 4 cams which are integral with the engine clutch shaft.

Both pumps, Fig. AC164, are of the 4 plunger type having 3 plungers of 11/16 inch diameter and one plunger of 5/16 inch diameter. Bolted to the top of the pump body is a subassembly called the hold position valve—and to the rear face is another subassembly which is the draft control. Hydraulic system on these tractors includes two rams, (a remote ram is available for trail type implements) a rockshaft and a load measuring spring providing draft control.

324. R & R AND OVERHAUL. On CA to remove pump and hold valve unit disconnect hand control linkage and hoses and unbolt pump and valve unit from pto housing. The hold valve may be detached from the pump **before** or **after** removing the pump from the pto.

On models WD and WD45 to remove the hydraulic pump, proceed as follows: Remove magnetic type drain plug from pump section of torque tube. Disconnect pump-to-ram hose

Fig. AC163—Model WD hydraulic lift pump installation on right side of tractor torque tube. Model WD45 is similar.

at right ram. Remove control housing-to-pump retaining cap screws and remove assembly. Remove pump-to-torque tube retaining bolts and nuts. Partially compress the pump cam followers and lift pump straight up.

After installing the unit, it will be necessary to bleed the system of air. Operate the pump and with both rams extended, loosen both ram packing gland nuts until a solid flow of oil is forced past the packing. Retighten packing gland nuts finger tight.

To remove pump cam shaft which on the CA is integral with the pto shaft refer to paragraphs 311 and 312. To remove camshaft which on the WD and WD45 is integral with the engine clutch shaft refer to paragraph 192.

CAM FOLLOWERS & PUMP PLUNGERS. After removing the pump, remove pivot pin to release cam follower and roller assemblies and pump plungers. Cam followers are available only as a complete assembly. Check outer surface of springs for wear and all over for rusting or corroded spots. Renew any doubtful springs.

HOLD POSITION VALVE. To disassemble the removed valve unit Fig. AC165, remove the snap ring (25) from camshaft (35), turn shaft to "lowering" position and withdraw shaft from body. Do not remove ball seat (31) unless it is known to be defective. If seat must be removed drill

and tap same for a screw puller.

With adjusting screw (C) turned all the way, valve should unseat at 875 to 975 pounds pressure. If valve leaks, try seating a new ball (1) to same by striking the ball a heavy solid blow using a one pound hammer and a soft drift interposed between ball and hammer. If necessary to install a new seat (31) reseat in same manner after seat insert has been assembled to valve body. If valve does not leak but releases at too low or too high a pressure vary the number of shims (W) to obtain desired 875-975 unseating pressure.

The point where valve body (30) is joined to pump body is sealed by an

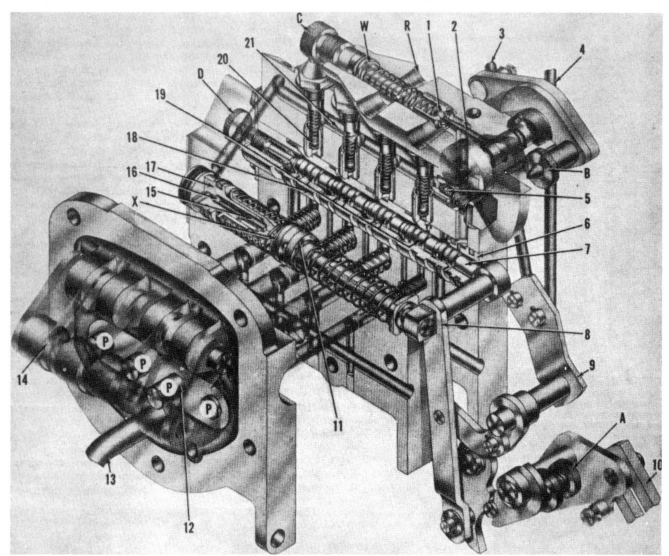

Fig. AC164—Model WD power lift hydraulic pump which is of the four plunger type. Pump is mounted on right side of torque tube. Models CA and WD45 are similar.

A. Draft regulating lock-out screw	1. Hold position valve ball	12. Pump pistons
B. Control lever lock screw	2. Hold position valve plunger	13. Oil inlet
C. Control valve adjusting screw	3. Control rod (pump)	14. Cam followers
D. Volume control adjusting screw	4. Control rod (hand)	15. Unloading valve plunger
P. Pump pistons	5. Check valve	16. Unloading valve ball
R. Right ram pressure connection	6. Lowering control valve sleeve	17. Unloading valve assembly
W. Hold position valve spring shims	7. Lowering control valve	18. Control valve sleeves
X. Unloading valve spring shims	8. Valve lever	19. Control valve plunger
	9. Control shaft	20. Discharge valve
	11. Unloading valve piston	21. Pressure manifold

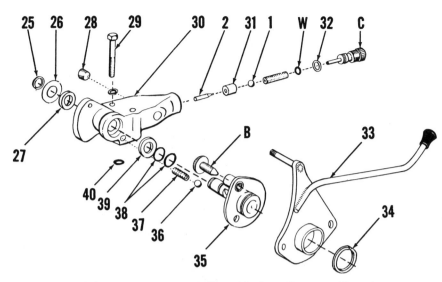

"O" ring (40). Similar rings are used on the camshaft. Always use new "O" rings when reassembling and reinstalling.

UNLOADING VALVE. To remove this valve from the pump body remove the screws from the cover (57—Fig. AC166) and remove cover. Carefully extract valve (54) from body with pliers. This valve is serviced as a selective fit assembly and if renewal is necessary it should be replaced as a unit. As an emergency measure a leaking valve can often be corrected by using a suitable press to push a new ball (16) against the seat at 5000 pounds pressure. If seat orifice measures less than 3/32 diameter after this procedure, it should be reamed to that size. End of plunger (15) which has the drilled hole should be assembled towards the ball. Valve should unload when pump pressure is 3400-3700 pounds. If it is positively known that valve does not leak, it can be adjusted to unload at desired 3400-3700 lbs. by varying shims (X).

Fig. AC165—Model CA hydraulic pump hold positioning valve assembly which is mounted on top of the pump unit. Models WD and WD45 are similar except for differences in the hand control linkage (33).

B. Control lever lock screw	2. Plunger	33. Hand control lever
C. Control valve adjusting screw	25. Snap ring	34. Snap ring
	26. Washer	35. Camshaft
W. Hold position valve spring shims	27. Dust seal	36. Detent ball
	29. Body retaining bolt	37. Detent spring
1. Hold position valve ball	30. Valve body	38. Seal ring
	31. Ball seat	39. Dust seal
	32. Seal ring	40. Valve body seal ring

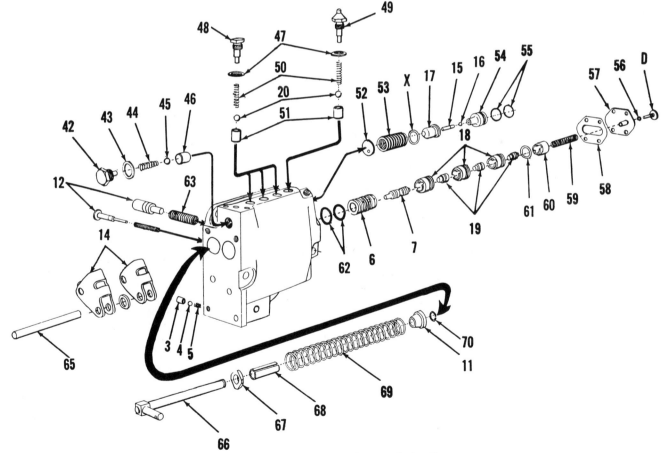

Fig. AC166—Models CA, WD and WD45 hydraulic pump.

D. Volume control adjusting screw	14. Cam followers	44. Spring	53. Spring	62. **Seal rings**
X. Unloading valve spring shims	15. Unloading valve plunger	45. Check valve (ball)	54. Unloading valve body	63. Plunger spring (11/16")
	16. Unloading valve ball	46. Check valve seat	55. Seal rings	64. Plunger spring (5/16")
6. Lowering control valve sleeve	17. Unloading valve assembly	47. Gasket	56. Seal ring	65. Cam follower pivot pin
	18. Control valve sleeves	48. Front discharge plug	57. Cover	66. Piston rod
7. Lowering control valve	19. Control valve plunger	49. Rear discharge plug	58. Cover gasket	67. Guide
11. Unloading valve piston	20. Discharge valve (ball)	50. Spring	59. Back-up spring	68. Spring stop spacer
12. Pump plungers (pistons)	42. Check valve plug	51. Discharge valve seat	60. Sleeve spacer	69. Linkage loading spring
	43. Gasket	52. Retainer	61. Spring washer	70. Snap ring

CONTROL VALVES. After removing the pump, remove control housing plate (57) and cover containing the volume adjusting screw (D). Remove valves (7 & 19) and sleeves (6 & 18) from rear of pump. Keep valves and sleeves in matched pairs. The valves for the ram and small (5/16) plunger form one unit which is sealed to the body by two "O" rings (62). Reinstall the valve sleeves and spacer (60) with notched ends of sleeves facing rearward. Spring washer (61) is installed between spacer and sleeve as shown in Fig. AC166. Be sure to use new "O" ring seals when reassembling the unit.

DISCHARGE VALVES. Seats (51) for the discharge valves should not be removed unless known to be defective. A leaky valve can often be corrected by reseating a new ball (20) to same by striking the ball a heavy blow using a one pound hammer and a soft drift interposed between ball and hammer. Valve seat inserts (51) can be removed by drilling and tapping the insert orifice to permit use of a puller screw.

DRAWBAR CONTROL UNIT. This unit, Fig. AC167, can be removed from the rear end face of the pump after removing the 4 bolts or screws which retain it to the pump. Method of disassembly is self evident after refer-

Fig. AC168—Model G hydraulic pump, and belt pulley unit drive housing assembly which is mounted on left side of engine clutch housing.

ring to the exploded view of the unit. Check cams and shafts and bushings for wear. Split type bushings may require final sizing after installation. For final sizing use a spirally fluted reamer or a hone.

325. **TEST OF PUMP AND VALVES.** In the absence of a test fixture for operating the plunger, the unit including hold valve and draft control can be tested by reinstalling it to the tractor. Pump should be primed before in-

Fig. AC169—Model G hydraulic pump unit gears when pump body is removed.

stalling by turning it upside down and pouring approved oil into inlet while working the plungers by hand. For temperatures above 10° F use SAE 20 or 20W oil. After running pump for a few minutes with control lever at bottom of quadrant bleed the system at ram packing nuts.

DELAYED LIFT. To test delayed lift valve first turn lockout screw (A—Fig. AC164) into pump body and lever screw (B) out of pump body. Turn adjusting screws (C) and (D) out. Connect a pressure gauge of 1000 pounds capacity at point (R) in series with the right (rear gang) ram. With the pump in operation, move hand control lever to top of quadrant. A gage pressure of 875-975 psi should be registered before the rear gang commences to move upward. To adjust the pressure, add or remove shims (W) located between adjusting screw (C) and spring.

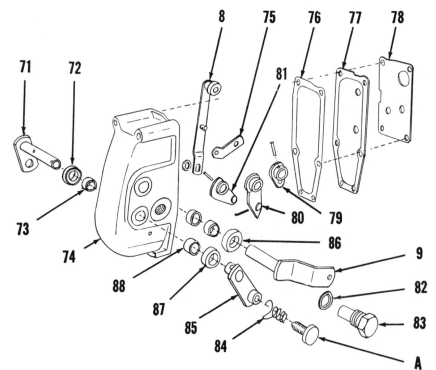

Fig. AC167—Model CA hydraulic pump drawbar control which is mounted on rear face of pump body. Model WD is similar.

A. Draft regulating lock-out screw	74. Control housing	80. Regulating cam	85. Lock-out arm
8. Control valve lever	75. Lever link	81. Regulating lever	86. Oil seal
71. Regulating shaft	76. Gasket	82. Gasket	87. Oil seal
72. Oil seal	77. Housing cover	83. Cam pivot screw	88. Bushing (long)
73. Bushing (short)	78. Pump body gasket	84. Lock-out screw spring	
	79. Hand control lever		

1. Bearing cover
2. Gasket
3. Shaft nut
4. Ball bearing
5. Shifter
6. Shifter pins
7. Drive shaft
8. Roller bearing
9. Housing
10. Shift arm
11. Shift lug
12. Snap ring
13. Belt pulley bevel
    drive gear
14. Seal ring
15. Hydraulic pump
    intake tube
16. Gasket
17. Bushing
18. Thrust washer
19. Needle bearing
20. Pump body
21. Pump center plate
22. Drive gear
23. Gasket
25. Pump gear (idler)
26. Seal
27. Shifter lever (to place
    BP in operation)
28. Detent spring
29. Gasket
30. Detent retainer

31. Detent ball
32. Needle bearing
33. Pump gear (idler) shaft
34. Pump gear (drive)
35. Dowel pin
36. Spring
37. Plug
38. Gasket
39. Retainer
40. Detent spring

41. Valve lever
42. Valve cover
43. Spring
44. Adjusting screw
45. Oil seal
46. Control valve
47. Valve cover stud
48. Ball seat
49. Ball
51. Detent ball

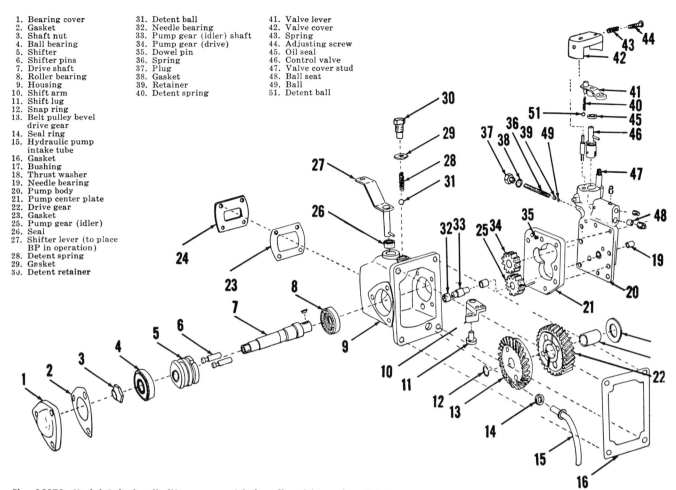

Fig. AC170—Model G hydraulic lift pump, and belt pulley drive unit which is mounted on left side of engine clutch housing. Cover (24) is used on housing (9) when unit is not equipped with a belt pulley.

UNLOADING VALVE. The pump unloads only when the rams are fully extended or the load is greater than the pressure required to unload the pump. To check and adjust the unloading valve (15 & 16—Fig. AC164); turn screw (A) out of pump body. Turn screw (B) into pump body separating the levers. Turn screws (C) and (D) out as far as possible. Connect a pressure gage of sufficient capacity (5000 lbs.) to the outlet side (R) of the pressure manifold (right rear ram pressure connection). With pump operating, move hand control lever downward. The valve should unload when the pressure is within the range of 3300-3700 psi. Adjust the unloading pressure by adding or removing shims (X) located between retainer and unloading valve spring.

326. TROUBLE SHOOTING. Causes for faulty pump or valve operation are outlined below:

DELAYED ACTION CANNOT BE OBTAINED. Screw (C) incorrectly adjusted. Check valve (5) not seating. Rams incorrectly connected. Binding implement and/or rams. Incorrectly

adjusted linkage between pump and hold valve (2).

IMPLEMENT RAISES BUT WILL NOT LOWER. Hold valve (1) not opening. Hold position link rod incorrectly adjusted. Binding implement and/or rams.

ERRATIC PUMP OPERATION. Incorrectly adjusted implement. Binding drawbar. Incorrectly adjusted pump-to-hand control link rod. Binding rams. Pressure line leak. Sticking control valves (7 & 19). Worn control linkage (8 & 9). Incorrectly adjusted pump-to-hold valve linkage (3). Leaking "O" rings. Sticking plunger or broken plunger springs.

LOW OIL PRESSURE. Unloading valve assembly (17) leaking at ball valve (16) and/or "O" sealing rings. Sticking control valve (7) in rear portion of pump. Pump plungers sticking or springs broken. Insufficient oil in pump reservoir.

HIGH OIL PRESSURE. Unloading valve seat orifice (17) restricted from ball hammering the seat or from dirt. Sticking control valves (19) in forward portion of pump.

Model G

The gear type pump Fig. AC168 is held to the combined hydraulic pump and belt pulley housing by cap screws. It can be serviced without removing the housing from the clutch housing. If the inner needle bearing for lower pump gear or the drive shaft is to be renewed it will be necessary to remove the pump and pump drive unit from the clutch housing.

328. PUMP OVERHAUL. Pump is removed by removing the 9 cap screws which retain the pump body (20). Be sure to use a suitable non hardening sealing compound on screw washers and underside of the head of each of the screws when reinstalling the pump. Refer to Fig. AC170.

With the body (cover) removed, the gears and their bearings are exposed for inspection as shown in Fig. AC169. End play of gears should be .002-.004. The mating faces of the center plate and pump body (cover) are lap finished to provide a gasketless joint. Avoid scratching or nicking of these surfaces. The pump gear shafts rotate in needle bearings some of which have

Fig. AC171—Model G hydraulic pump and belt pulley drive unit when viewed from the mounting pad side.

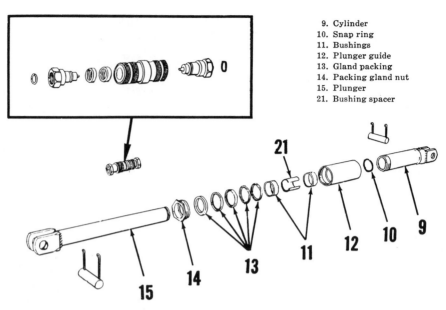

9. Cylinder
10. Snap ring
11. Bushings
12. Plunger guide
13. Gland packing
14. Packing gland nut
15. Plunger
21. Bushing spacer

Fig. AC173—Model CA remote hydraulic ram—exploded view. Install plunger gland packing (13) so that the chevron ring lips face toward the pressure.

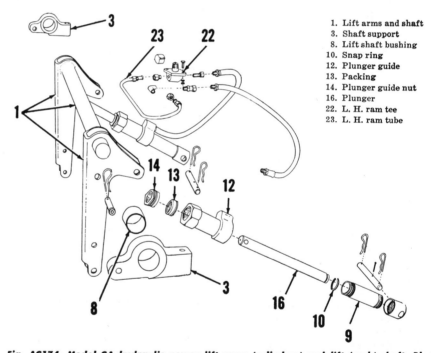

1. Lift arms and shaft
3. Shaft support
8. Lift shaft bushing
10. Snap ring
12. Plunger guide
13. Packing
14. Plunger guide nut
16. Plunger
22. L. H. ram tee
23. L. H. ram tube

Fig. AC174—Model CA hydraulic power lift rams (cylinders) and lift (rock) shaft. Bleed hydraulic system by loosening packing gland nut (14) when the rams are fully extended.

closed end cages as shown at (19). To insure against fluid leakage each closed end bearing cage should be lightly coated with an approved gasket cement when it is being installed. Refer to Fig. AC170.

Pump relief valves and control valve are mounted in top of pump body as shown. Do not remove ball seat (48) unless it is known to be defective. A leaking valve can often be corrected by striking a new ball (49) against the seat using a one pound hammer and a soft drift. Relief valve spring (36) has a free length of 3 13/16. The screw (44) and spring (43) in control cover is used to adjust the rate of implement drop. If it is turned clockwise, the drop rate is decreased. A valve in the remote ram line adjusts the amount of delayed lift of the rear furrowing bar.

When tested with a pressure gauge, the pump relief valve should unload at 900 psi.

To overhaul the pump drive shaft, (7), which also drives the belt pulley refer to paragraph 309B.

HYDRAULIC RAMS

329. Method of removal and disassembly of these units, Figs. AC173, AC174, and AC176 is self evident. On some models the ram plungers act simply as elements to displace the fluid, and therefore, they are considerably smaller in diameter than the bore of the unmachined cylinders in which they operate. It is advisable to renew "O" or chevron packings whenever a unit is disassembled. On models having chevron type packings the packing should be installed with the open end of the "V" towards the oil supply. Reject any plungers on which

the seal contacting surfaces are scratched or scored. Rams and cylinders must be bled to remove all air from hydraulic system after they have been installed.

HYDRAULIC SYSTEM LINKAGE

Because of the draft control characteristics of the hydraulic system on models CA, WD, and WD45, it is important that the system linkage be correctly adjusted as outlined in paragraphs 330 through 336.

Model CA

330. **PUMP REGULATING SPRING.** The recommneded adjustment is ap-

proximately 3/16 of an inch of preload as shown in Fig. AC177. This can be obtained by backing off the nut (11) until spring is just free endwise, then tightening the nut until the spring has been shortened 3/16 inch which is the equivalent of 3 complete revolutions of the nut.

331. **DRAWBAR LINKAGE.** This linkage, Fig. AC178, is correctly adjusted when with zero load on the drawbar and drawbar clamp loose there is no clearance at (4) and 1/32 inch clearance at (X) when hand control lever is in raised position.

1. Lift arm
2. Lift shaft
3. Shaft bracket
4. Left side ram hose
5. Pump adapter
6. Right side ram hose
7. Ram support
8. Ram bottom cap
9. Ram cylinder
10. Snap ring
11. Bushing
12. Guide
13. Packing
14. Gland nut
15. Plunger
16. Snap ring
17. L. H. & R. H. crank
18. Bushing
19. "O" ring seal
20. Swivel connection

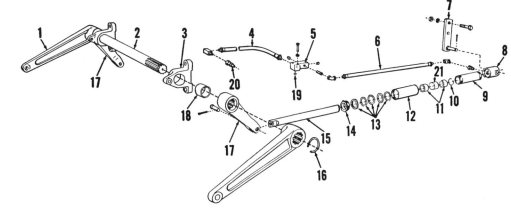

Fig. AC176—Models WD and WD45 hydraulic power lift ram (cylinder) and lift (rock) shaft. Bleed hydraulic system by loosening packing gland nut (14) when the rams are fully extended.

Method of adjustment is as follows:

Lock out drawbar control by tightening the lockout screw (A) in lockout arm on pump, loosen drawbar clamp and make sure pump regulating spring has correct 3/16 inch preload as in preceding paragraph, 330. Loosen lock nut (8) and lock nut (6) and back off nut (5) to permit rod (2) to bottom in arm (1). Now turn nut (5) in opposite direction until it brings washer (4) into contact with bracket

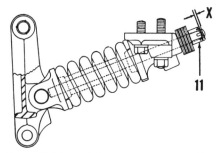

Fig. AC177—Model CA pump regulating spring adjustment is made with nut (11) so as to preload the spring approximately 3/16 inch.

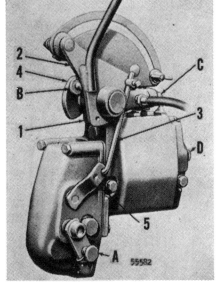

Fig. AC179—Model CA hydraulic lift pump hand control lever. Refer to paragraph 332 for adjustment procedure.

A. Draft regulating lock-out screw
B. Control lever lock screw
C. Control valve adjusting screw
D. Volume control adjusting screw

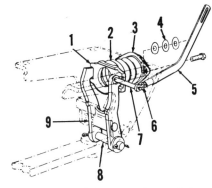

Fig. AC180—Models WD and WD45 drawbar control mechanism.

1. Spring seat	6. Link rod adjustment
2. Spring	7. Link rod stud
3. Spring retainer	8. Fork pin
4. Shims	9. Control fork
5. Drawbar link rod	

(3). Relock the lock nut (6). With hand control lever in "raised" position, turn screw (9) until a 1/32 inch clearance (X) is obtained between top of shank (10) and underside of slotted head of screw (9).

332. HAND LEVER LINK. The purpose of this adjustment is to establish the correct length of the link rod which connects the hand control lever to the lever on the pump control housing. Turn screw (D—Fig. AC179) out until it contacts the collar on shifter lever shaft. Lock the drawbar by turning screw (A) into control housing. Lock plates (1) and (2) together with screw (B). Loosen set screw, freeing link rod (3) and place hand control lever in down position with point of screw (B) aligned with upper edge of hole (4) in valve body. Now rotate lever (5) downward as far as it will go and tighten the set screw at upper end of link rod (3). Turn screw (C) out until it stops against plug.

Models WD-WD45

334. PUMP REGULATING SPRING. The preload of this spring (2—Fig. AC180) is set at the factory by vary-

Fig. AC178—Model CA hydraulic system drawbar linkage. Refer to paragraph 331 for adjustment procedure.

X. 1/32 inch **clearance**
1. Drawbar support **arm**
2. Regulating rod
3. Drawbar bracket
9. Adjusting screw
10. Link

ing the number of shims (4) and should require no further attention. Always reinstall same shims as were removed.

335. DRAWBAR LINK ROD. Length of link rod (5—Fig. AC181) should be such that it will just fit over connecting pin when the drawbar is free in the forward position and draft regulating lockout screw (A—Fig. AC164) is turned into housing. Too long a link rod will decrease the draft, and if rod is too short it may break the control linkage.

336. HAND CONTROL ROD. To adjust the link rod which connects the lever on steering column to the lever on pump, first loosen the set screw at upper end of rod. Place hand lever at top of quadrant and push control lever on pump down. Tighten the set screw.

MECHANICAL LIFT

Models RC-WC

340. To disassemble the power lift, remove the unit from the tractor. Remove pin (18—Fig. AC182) from worm gear (2) and drive power lift shaft (15) out far enough to remove pin (5) from ratchet (6). The need and procedure for further disassembly will be determined by an inspection of the parts and by reference to Fig. AC182. When reinstalling unit on tractor, vary number of gaskets between housing and torque tube to provide a slight amount of gear backlash and install sufficient shims (12) between housing and frame to assure correct alignment.

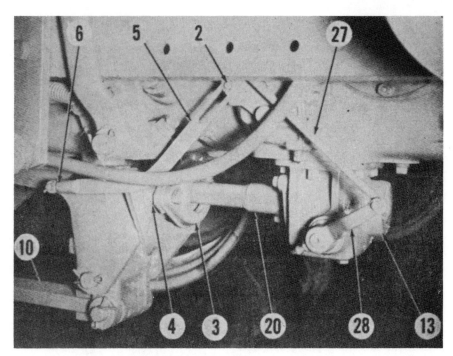

Fig. AC181—Model WD power take-off drive housing installation and front view of drawbar control. Model WD45 is similar.

2. Draft control lever pin	4. Spring shims	10. Drawbar	27. PTO hand control
3. Spring retainer	5. Drawbar link rod	13. PTO drive housing	28. Hand control shift lever
	6. Rod alignment	20. PTO extension shaft	

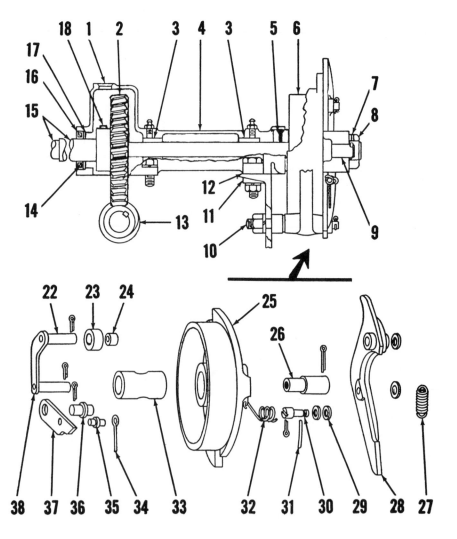

Fig. AC182—Models RC, and WC mechanical type power lift unit which is mounted on top of the torque tube.

1. Expansion plug	18. Worm gear pin
2. Worm gear	22. Roller arm pin
3. Shaft bushing (effective 1941)	23. Trip roller
	24. Roller bushing
4. Power lift housing	25. Cam
5. Ratchet pin	26. Cam crankpin
6. Ratchet wheel	27. Trip spring
7. Shaft collar pin	28. Trip
8. Shaft collar	29. Shim washer
9. Cam crank pin	30. Fulcrum pin
11. Bevel washer	31. Pawl shifter
12. Aligning shim	32. Pawl spring
13. Power lift worm (in torque tube)	33. Cam bushing
	34. Pawl plunger
14. Shaft oil seal	35. Pawl pin
15. Power lift shaft	36. Pivot pin
16. Oil seal retainer	37. Pawl
17. Oil seal washer	38. Trip roller arm